For

Live in harmony

♡

IN THE
FOOTSTEPS
OF
RACHEL
CARSON
HARNESSING EARTH'S
HEALING POWER

PATRICIA M. DEMARCO, PH.D.

Patricia M. De Marco

urbanpress

Jan. 4, 2023

I leave this book 3/23
on the final day of my cruise.
The author is a wise woman
and a friend.
Listen to her, learn from her
& pass it on!
Connie Fortis, Pittsburgh

In the Footsteps of Rachel Carson
by Patricia M. DeMarco
Copyright ©2022 Patricia M. DeMarco

ISBN 978-1-63360-203-8

For Worldwide Distribution
Printed in the USA

Urban Press
P.O. Box 8881
Pittsburgh, PA 15221-0881
412.646.2780

This book is dedicated
to the children of the 21st century
with hope they will thrive in a healthy world,
in honor of Rachel Carson.

CONTENTS

Photo credit: Jennifer Claudine Smith

FOREWORD

Patty DeMarco is an exemplary model of "the power of joined voices." Her voice as heard through her Rachel Carson Legacy symposia, a documentary on Rachel Carson, and her work and writing all bear witness to the belief that all of earth is intricately and beautifully interwoven in the web of life. Indeed, she follows in the footsteps of Rachel Carson as an activist and visionary for the recuperating future of the earth.

Rachel's first footsteps were in an old farmhouse in Springdale, Pennsylvania, not far northeast of the sooty, stifling industrial center of Pittsburgh. I have been one in a long line over the past 50 years of presidents of the Rachel Carson Homestead Board. From early childhood, Rachel's educated and education-minded mother, the daughter of a United Presbyterian minister, walked with her through the woods of the 65 acres her father had purchased as a real estate investment. Like Patty watching the widgeons on Oney Pond in Alaska, Rachel was acquainted with the songs and identification of birds. One of Rachel's early published writings was at the age of 15, "My Favorite Recreation." She heard the "cheery 'witchery, witchery' of the Maryland yellowthroat" (now called the Common Yellowthroat), and the story was published in *St. Nicholas* magazine. Her exploration of and engagement with all of creation became a life-long pursuit and delight.

Patty's devotion and defense of the environment led her to Rachel's Homestead. As executive director, she revived/reimagined/continued Rachel's unrelenting, outspoken voice. Patty coordinated and spearheaded eight Rachel Carson Legacy Conferences that gave momentum and agency to Rachel Carson's legacy. Inviting voices from around the country who were in the vanguard of research on endocrine disruptors and green chemistry, alternatives to fracking and celebration of biodiversity with Edward O Wilson, these multidisciplinary conferences crossed the academic silos and engaged a wide community of interests. A thousand people gathered to dedicate the Rachel Carson Bridge in 1906 over the water her example inspired in another Springdale native, Michael Watts, who helped clean up the Allegheny River, a microcosm of Earth Day.

Patty's footsteps also coincided with Rachel's through frightening, formidable valleys. The darkness of tumors and cancer recurred through decades of each of their lives. Doctors concealed the truth from Rachel repeatedly, a common practice in the male-dominated profession. Even though separated by half a century, their experiences were similarly bleak and foreboding. Each found invigoration during her illness in the quest and crusade to educate the people and assault the perpetrators, hopefully leading to a seismic shift from exploitation to nurture of all of creation!

May we all be inspired and motivated to action. May we learn from Rachel Carson's sense of wonder and the comprehensions of the nurture and attacks on the earth which began for her in the woods of Springdale. Join the advocacy and action of the Rachel Carson Council that succeeded her. Seek out the "the bonds formed with the land there [which] have sustained me all of my life." May we all reach for this depth that Patty has shared: "I hold within me to this day the impetus to move in harmony with Nature, to express my deepest emotions, joy, sorrow, fear, anger and longing in resonance with movement in natural spaces."

Rev. David Carlisle
President, Rachel Carson Homestead Association
April 2022

INTRODUCTION

Rachel Carson's legacy is difficult to measure. She wrote *Silent Spring* while dying of metastatic breast cancer and a host of other ailments. Yet numerous idealistic and impressive high school students still find her a fascinating figure. They call to ask me for their National History Day projects, "What was Rachel Carson's impact? What were her main achievements?" I am not joking when I tell them an entire book would be needed to answer that. Yet, the banning of DDT; the saving of the eagle, osprey, and Peregrine falcon; the Clean Air Act; the Endangered Species Act; the EPA; Earth Day; the eventual linking of breast cancer and other diseases to the environment were all achieved by others after Carson's death in 1964.

And so, I tell these young women, and many others, that through her work, her writing, and her life, Rachel Carson is an eternal inspiration. Her legacy depends on what each of us does when we come upon Rachel's gifts—imagination, awe, wonder for nature and for life itself; empathy for all living creatures, no matter how distant, or seemingly different from ourselves; and brave action and advocacy in the face of relentless attacks, opposition, and ridicule. Carson said, "Knowing what I do, I could not remain silent and still live with myself."

Can we measure inspiration? Most of the evidence would of necessity be anecdotal. I come across many individuals whose own effect on society and those around them is impossible to measure, who tell me that Rachel Carson touched their lives, taught them to care about others and the environment, and to carry on. And then there is direct evidence, testimony in word and deed, from those in public life who describe Rachel Carson as a gift that has guided their life's work—such as former Vice President Al Gore, writers like Sandra Steingraber or Terry Tempest Williams, historians like Linda Lear or Douglas Brinkley, marine biologists like Sylvia Earle, heads of federal agencies and environmental groups like Jamie Rapaport Clark, and many more.

With *In the Footsteps of Rachel Carson*, we have compelling evidence of the powerful gravitational pull of Rachel Carson's legacy on today's environmental leaders and its ripples that have given us Patty DeMarco. DeMarco is the first direct heir to Rachel Carson's legacy who has written a refreshingly revealing and honest memoir blended with reflections on today's difficult and despairing times. DeMarco's parallels with Carson's life are numerous—born and raised working poor in Pittsburgh, connected to Chatham University, trained in graduate biology, positions as a part-time academic, not a tenured "expert," and a professional life interrupted by difficult family issues and painful battles with cancer.

But DeMarco's childhood exposure to the natural world is not the woodland walks of Rachel Carson, but a backyard in the Mt. Washington neighborhood designed to feed a large family—pear, plum, and cherry trees, onions, tomatoes, zucchini, beans, spices growing between brick walkways, a replica of the Italian countryside, including fig trees brought by her grandfather from his hometown of Campolieto, Campobasso. "Pop" was a skilled stone cutter for the Pennsylvania Railroad, but DeMarco did not feel poor, surrounded by the scents and sumptuous food the family canned, pickled, and packed in salt and pepper. There were even illegal chickens under the porch for eggs and chicken cacciatore. This feasting on what we would now call local, organic, and fresh food was also a community affair as many families in the neighborhood created such urban oases and shared and worked together under the shade of the grape arbor, or in the kitchen, chatting away in Italian.

DeMarco credits her observation and love of even the tiny things in nature—plants, flowers, insects—to Pasqualina, her grandmother, her Nona. From her mother, Marcella, a pioneer in women's power, she learned to stand up for those treated unjustly and to speak out against the scourge of war, including the bombing of Hiroshima and Nagasaki. Like Carson, Patty DeMarco also has a wonderful sense of humor beneath earnest entreaties. You can hear the chuckles as she writes of her Uncle Louie. He is loud, reeks of garlic, which he eats raw for breakfast, and uses Patty's love of the little things in nature to pick blossoms of dozens and dozens of dandelions. These are promptly fermented into dandelion wine, which mixed with garlic breath, creates odors that get choice seats on the streetcar for his nieces.

DeMarco's *joie de vivre*, sense of community, caring for others, and love of nature are, again like Carson, blended with serious science first gained from what we now call experiential learning that suffuses the recollections of her youth. As the immigrant DeMarco family moves up (Mom became a teacher and Father worked at the Buhl Planetarium and later became a diplomat), young Patty DeMarco marvels at the Sky Show, her father's deep, resonant narration ("Long ago and far away the Milky Way was formed"). Patty is allowed to help with the static electricity demonstration, proudly letting a wand be waved over her head to make her hair stand on end.

Later, DeMarco's father is stationed in South America where she emerges from childhood into a teen who finds her independence and confidence, her calling as a student of biodiversity and social justice, and her place, her sense of belonging—her center. DeMarco's description of her time in Rio de Janeiro near an isolated fishing village in a cliffside house around the curve from Copacabana on the edge of both jungle and poor *favelas* is a masterpiece reminiscent of Von Humboldt and redolent with aromas and rich colors—rotting logs, cascading small yellow and brown orchids, and iridescent blue butterflies. At eleven, Patty examines the cell patterns of butterfly eggs, dead beetle legs, or leaf layers with her Fisher Price microscope, listens to monkeys in her tree house as she sits looking at the ocean, and, on the beach, watches fisherman haul in skates, or lose an octopus as a wave washes it away.

Since DeMarco's father chose not to have his family live in a high rise within the American Compound, but picked instead a rural place with tiered gardens, vegetables, and fruit trees that resemble Italy, Patty is both independent and isolated in her teens. Her classmates in Our Lady of Fatima think her "distinctly weird," bookish, shy, and with a strange accent. This highly unusual adolescence for an American girl is much like that of Rachel Carson who stayed close with her mother, Maria, did not date, and found strength in books, museums, and the study of nature.

All this, for DeMarco, is preparation for adulthood, much of it married with children in Connecticut, where she frequents Hammonasset Beach State Park and, instead of lying on a towel under an umbrella "like other Proper Mothers," channels Rachel Carson, combing the beach for shells and egg cases, squatting over tidal pools with a magnifying glass, watching hermit crabs under stone ledges, and peering at mussels and barnacles clinging to the rocks. With her son Steve, she walks the wetlands, worrying about the lives of the razor clams they look at, returning every specimen, as did Carson, to its home.

But like Rachel Carson, whose graduate study toward a Ph.D. in biology at Johns Hopkins was interspersed with half-time work, teaching adjunct courses, difficulties in nailing down an adequate research topic, and the need to provide for her extended family, there are obstacles. Patty DeMarco's graduate work for a Ph.D. at the University of Pittsburgh requires her to move into her old home and be away from her husband while he finishes his medical internship in Connecticut. Other wives are horrified she is not supporting her husband, while her family is shocked that she has not yet had children (she is on that early pillar of women's liberation, the Pill). Her Auntie is even convinced she has had an abortion since her wedding date was moved up from August to April. DeMarco spends endless hours dutifully studying *Drosophila* larvae, but her heart is in helping canoeing friends who collect Mason jars full of water from the Monongahela River polluted by runoff from the steel mills. Patty checks for mutagenicity and wants to do her dissertation on it. Her advisers tell her starchily to stick to a topic with a well-established literature base. There just isn't enough published research on environmental mutagens.

DeMarco finishes her Ph.D., but is later divorced from her doctor husband and has a startup firm she has launched with two friends and a $5 million ARPA grant taken over by politically-connected men displaced from the shutdown of Electric Boat in Groton, Connecticut. She heads off to Alaska with an unemployed General Motors machinist with a motorcycle who aspires to be a fishing guide.

In Alaska, with women as lieutenant governor, president of the senate, speaker of the house, and many other strong women, including matriarchal Alaska native communities with female elders, DeMarco flourishes. She is at the peak of her career, leading the Alaska Economic Development Corporation, serving as a commissioner of the Regulatory Commission of Alaska, and associate dean of the College of Business and Public Policy at the University of Alaska, Anchorage when cancer strikes.

As with Rachel Carson's breast cancer, DeMarco conceals her uterine and fallopian carcinomas except with close friends since she is participating in a review of the Trans-Alaskan Pipeline Tariff and does not want her work undermined by oil companies. How does one carry on with such a life-threatening diagnosis and the agonies of chemotherapy? The words from Rachel Carson's *Silent Spring* were often quoted: "Those who contemplate the beauty of the earth find reserves of strength that will endure as long as life lasts." After breast cancer surgery, Carson sought solace on the beaches of Nags Head, North Carolina. More restrained and private, Rachel Carson does not tell us how she found reserves of strength, nor does she connect the beauty of the dunes and dappled waves with her cancer or facing death. Patty DeMarco does.

Her Alaskan "Scenes from Oney Circle," where she has a log house near the Chugach National Forest with a view of the Anchorage Bowl and Denali, lets us see, feel, and wonder at the healing power of nature. In her driveway, walking past alder bushes glittering with ice, with Orion and Cassiopeia overhead, DeMarco finds two young moose bedded down behind her house, curled together, their body heat melting a cave in the snow to protect them from the wind. It is Christmas and this scene is suddenly illuminated by the aurora borealis, the northern lights, as "ribbons of green, pink, yellow cascaded down toward the horizon like lightning in slow motion." There are more than seventy-five ducklings she watches in her pond—a mallard mother with a lame leg; pintails; a lone Barrows Goldeneye, whose nest has been raided of her chicks by a great horned owl; and widgeons "with their taupe feathers and beautiful dark eyes." There are eagles, close encounters with moose and bear, seals fishing in the water, and an epiphany following her cancer surgery.

DeMarco is amidst the alders by her pond wrapped in an afghan with a cup of hot tea, in the fog of pain killers, listening to chickadees, and watching an ermine turning white sneaking around her dock. A mother lynx, wary and alert, creeps stealthily forward to lap the water, keeping her eyes lifted while she drinks. Patty can only see the brush rustle where her cubs are hiding and waiting. Later, she finds lynx tracks in the snow, but those of just one cub. The trials of this mother stay with her throughout her treatments. The sound and sight of this mother lynx lapping up the water of life creates a source of strength and still point for DeMarco, one to which she often returns as she struggles with her own surgery and sorrow of which there is more to come. She develops breast cancer late in life and again dances with the "red devil" of chemotherapy and undergoes a bilateral radical mastectomy. Worse, this is at the exact time she gets the news that her daughter also has breast cancer.

Today, DeMarco's reserves of strength are still sorely needed, but not just those derived from enjoying or contemplating nature. The spread of chemicals and the crisis of climate change threaten all of us more than even Rachel Carson might have imagined. By this time, Patty DeMarco has become a symbol of Rachel Carson's legacy. She has headed the Rachel Carson Homestead in Springdale, chaired the Rachel Carson Institute at Chatham, taught about Rachel Carson, and fought for the environment in Pittsburgh—her home and the root of her compassion laced with a resilient, joyful spirit.

But in the face of global existential threats, it is DeMarco's defiance that marks her as an heir of Rachel Carson. Both women have been made strong by adversity, both displaying unwavering empathy and compassion, both loving life and all of nature so fiercely that they must speak out. Carson said, "Conservation is a cause without an end." And so, DeMarco says, "Fighting to replace the fossil industries feels like a direct confrontation with the probable causes of my many tumors.... I live on borrowed time but stand with defiance in the face of the challenges both to my own health and to the health of our society." Patty DeMarco can still speak to and reach mainstream America, but now she rouses crowds at Earth Day, stands in protest with indigenous people fighting against pipelines, fracking in Pennsylvania, or nuclear power and weapons whose radiation should, instead, be used to battle cancer.

We cannot carefully chart the course of how our words or actions matter. We cannot fully know how smart and strong women find reserves of strength in interrupted and obstacle-laden lives. Nor can we measure precisely the influence, or the inspiration, of Rachel Carson. But there is evidence enough in Patty DeMarco's *In the Footsteps of Rachel Carson*, from her colorful and caring childhood in the Italian section of Pittsburgh, chatting under the grape arbor; or in the kitchen, with a community of women and her Nona, Pasqualina; to the empathy and identification with a mother lynx; to the proud and powerful voice of an environmental elder who will not be silent.

Robert K. Musil, Ph.D., M.P.H.
President, Rachel Carson Council
Bethesda, Maryland

Robert K. Musil, PhD, MPH is the President and CEO of the Rachel Carson Council, the legacy organization envisioned by Rachel Carson and founded in 1965 by her closest friends and colleagues. Dr. Musil was named President and CEO in February, 2014 and is only the third head of this historic environmental group.

PROLOGUE

I have had four bouts with cancer between 2001 and 2018, each time losing parts of my body—each time gaining resolve to defy my fate. Nobody really expects to be diagnosed with cancer until it happens. The disease, in whatever manifestation it takes, wreaks havoc with one's life. One in eight women in America will have breast cancer. While it is not the scourge it once was when Rachel Carson suffered mostly in private, hiding her illness from her colleagues and her public, each individual copes with cancer as a personal battle for survival.

Rachel Carson has been my beacon and inspiration. My life experience parallels hers in many ways, and at two critical intersections, her writing gave me direction and focus. I shared her experience of connecting closely to nature as a child, working as a scientist in the public policy arena, and having cancer while holding a public position.

There are many ways people face cancer, and no one prescription works for everybody. I share my own journey in this book out of the conviction that to have healthy people, we need a healthy world. Connecting to the universal healing energy of the living earth is the strongest force for survival I know—universal across religions, cultures, races, genders, and generations. Connecting to the power of Mother Earth through nature brought me healing and hope in the face of a cancer diagnosis.

I write this book as a tribute to the life force of women—the progenitors of culture, the tenders of community, the givers of life. Defiance defined the women of the "liberated" generation—defiance against the patriarchal traditions and the constricted pattern of women's place subservient to men. It is also defiance to the attack of cancer that takes our physical female attributes of womb and breasts but cannot take our soul, our smile, or our determination. For me, the connection to the earth has been my source of strength because everywhere I see the resilience and beauty of nature. This bond with nature as a source of solace and resilience I share with Rachel Carson. As Director of the Rachel Carson Homestead Association, I delved deeply into her life story and was saddened but also inspired to know that she was suffering from breast cancer during the whole time she wrote *Silent Spring*.

My defiance toward the forces that perpetuate pollution, corruption, repression and even disease is rarely drawn from rage and anger. As with Rachel Carson, I engaged in a series of battles using arguments based on science to combat institutional barriers to living in harmony with nature. Connecting to nature through daily quiet meditation while walking, journaling, or just sitting in a tree has always offered me a way to escape painful realities and to find courage to persist in the face of constant obstacles.

While I was going through cancer treatments, deliberately killing part of myself to exterminate the tumor cells through chemotherapy and surgery, I did not speak of my misery. I maintained an overtly positive outlook and demeanor to those around me as much as possible. I refused to give cancer my words. I used my journal to scream out my frustration, pain, confusion, and fear while I also reflected on my observations from nature that helped me for a moment to escape my own suffering. Nature, whether minute and intimate or sweeping and grand, offers daily opportunities for joy in and testimony to the robust resilience of life. From this process of writing and reflection, I drew strength for the internal battle.

In my public life, I have used the power of positive examples of success to inspire change. Mine has not been the defiance of striding through the streets in rage. Rather, I have infiltrated the institutions of government and education to drive change in laws, to inform policies in favor of protecting the environment, and to inspire people to visualize a better future. People will not move toward an unknown; they must be able to visualize a new situation and see it as an improvement.

The first section of this book lays down the foundation of family traditions as a source of strength as well as the standard for rebellion against adversity, whether the battle is internal as the battle with cancer or external as the battle against the pollution of the earth. The second and third sections draw from my experiences in nature as I extended my engagement to the workplace, then to public institutions, always operating from my conviction that we cannot have healthy people in a sick world. The fourth section reflects on my journey and a commitment to maintain a positive direction going forward from the perspective of my public life as an elected official. Having cancer sharpened the focus of my life—a determination to live with purpose every day, to live in harmony with nature. I hope that sharing these essays will bring solace and strength to you when you must travel this path, sometimes struggling to thrive in the face of the challenges of our time.

In this book, I trace my own path in the footsteps of Rachel Carson from my childhood bonding with nature, through my rebellion toward the patriarchal restrictions of my cultural heritage, and on to the defiance of laws and civil actions that enrich corporations but destroy nature and dehumanize communities. This is my journey—a defiance of disease that would shut me down, and a triumphant tribute to Rachel Carson who blazed the path so long ago.

Rachel Carson believed that those who know have the obligation to speak up in defense of nature. I have taken that charge seriously in these times of existential crises. When suffering from a potentially life-threatening illness, it is tempting to curl in to oneself and close out the demands of the wider world. It is excusable to isolate and turn inwards, and for many cancer sufferers, this is a path of solace. But for me, my path is to defy the limitations of the illness for as long as possible.

Cancer has forced me to make choices, to parse out my strength for things that matter most. Standing at the head of marches with inspiring words can happen, but the actual marching, I must forgo. Giving exhortations to action on the radio or in a written post invigorate me, but the actual canvassing door to door, up and down the hills of Pittsburgh must fall to younger colleagues. I realize that as Rachel Carson wrote *Silent Spring* in spite of her debilitating experience with breast cancer treatments, she was shouting defiance to the industrial complex emerging at that time to infuse our world with pesticides, herbicides and synthetic materials. I follow in the footsteps of her battle, harnessing the power of joined voices to build a better world, a future of shared prosperity infused with purpose and meaning, a future where we can live in harmony with nature.

Patricia M. DeMarco, Ph.D.

PART I

ROOTED IN THE EARTH

The DeMarco fig tree

*Those who contemplate the beauty of the earth
find reserves of strength that will endure as long as life lasts*
– Rachel Carson (*The Sense of Wonder*, p. 100).

Part One

     I was born in 1946 in Pittsburgh, PA where my family lived upstairs from my father's parents until I was five years old. Then, my father joined the U.S. Foreign Service and we traveled all over the world for postings, two or three years at a time, in Bogota, Columbia; Manila in the Philippines; and Rio de Janeiro, Brazil. We returned to Pittsburgh as I entered high school, and I attended Dormont High School. Then I matriculated at the University of Pittsburgh for undergraduate and doctoral degrees majoring in biology with my thesis in genetics.

     I married my husband, Gordon Smith, in 1968 while he was attending the University of Pittsburgh Medical School and I was doing my graduate work in genetics at Pitt. After our graduate studies, we settled in Cheshire, Connecticut where he set up his medical practice. We had two children and my struggle to be a wife and mother while maintaining a professional career finally culminated in divorce in 1992 when both children had gone away to college.

     This first series of essays shows how and why I became grounded in nature as my touchstone. It frames the source of my philosophy around governance by the laws of nature and defines the importance of family ties. This part also frames my struggle with traditional roles of wife and mother and my urge to make a difference in the world through my own professional actions. Like Rachel Carson, my family ties were formative of my world experience. Also like Rachel Carson, obligations to family shaped my experiences and opportunities, as the eldest daughter helping to care for the family so Mother could work.

1.1 WALKING IN RACHEL CARSON'S FOOTSTEPS

I first encountered Rachel Carson's writing when traveling on the ship from Brazil back to New York in 1958. A copy of *The Sea Around Us* was on the coffee table in the ship's lounge. I was fascinated to read it while traveling on the ocean experiencing the power of the seas. I received a copy of *Silent Spring* for my high school graduation present in May 1964, and reading her work inspired me to pursue a career in biology, with a special interest in genetics.

At times, I identified quite closely with her journey. In the summer of 2010, I visited the Chincoteague National Wildlife Refuge in the company of Lou Hines just before he retired as the director. I was visiting as part of the research for my first book, *Pathways to Our Sustainable Future*, and wanted to experience Chincoteague from Rachel Carson's perspective. He took a small Jeep and we drove to the old dock entrance to the refuge where people would have come ashore before the bridge was built connecting Assateague Island to the mainland.

Lou took me down a rough road used only by maintenance crews. We were well swathed in protective hat netting, long sleeves, boots, and gloves to protect us from the swarms of mosquitoes and greenflies. As we moved through the ecotones from the woods to the seashore, I had an acute sense that Rachel Carson had walked this way years ago, when I was just a toddler learning to walk in Nona's garden. When we reached the edge of the bay, I walked alone down the packed sand, thinking about the path Rachel Carson had traveled in her life, and the many ways my own path intersected the same space, but at a different time. This is the magic of preserved spaces:

the experiences and sensations shared across time. That unmistakable scent of salt marsh, sea spray, and wind coming over open water evokes memories shared across generations. This sense of a shared purpose and destiny was reinforced by the disease we shared and fought.

As I traveled in Rachel Carson's footsteps for much of my life, I became even more aware of her courage as she faced the devastation of cancer in the face of so much adversity. When she was writing *Silent Spring*, she was dying of breast cancer, undiagnosed and untreated until very late in the progression of the disease. I often wondered how such a thing could happen. Then I realized that in her time, the doctors were nearly all men, and the societal norms did not support single women in their dialogue with doctors about female issues like breast cancer. Even in the early '70s, my own discussions with my gynecologist about having birth control pills so I could finish my doctorate before having children required assuring him that my husband did not object.

Rachel was undergoing radiation treatments as well as a mastectomy of her left breast during the time she was working on *Silent Spring*. Her battle with the disease was made more difficult not only by the limitations to her work from exhaustion, iritis, and pain but also from the need to avoid public knowledge of her illness lest her detractors use it to discredit her work.[1]

Rachel wrote a letter to her physician in which she challenged him to tell her the truth about her illness. In it she noted that the doctor's wife by the name of Jane had been a beacon to her since Jane was also battling breast cancer. But after Jane died, Rachel felt that she had lost her guide. She documented swellings she found in her neck and lymph nodes along with increasing pain in her shoulder, back, and legs, and noted, "At the time I went (to the radiation doctor) about my back in December I kept making remarks about having 'arthritis' in my left shoulder, but no one paid much attention."[2] She demanded to know whether these symptoms signaled a more advanced progression of her disease so she would have time to tidy up important matters. This entry was just months before her death in April 1963.

I thought of Rachel Carson often as I was undergoing cancer treatments, thankful for the advances in medicine due to increased understanding of the underlying basis of the disease that allowed much improved treatments. Even in the span of time between my first and second cancers, much was done to control side effects and limit damage to essential functions. I marveled that Rachel Carson had kept her spirit to complete her work even knowing that her days were very few.

I often examined Rachel Carson's effectiveness with my students and audiences when representing the Rachel Carson Homestead and the Rachel Carson Institute at Chatham University. The power of her lone voice to move Congress inspires me and many others to this day. Her credibility came from years of research and documentation of the oceans and the ecology of estuarine areas. Her influence came from her ability to bring the science to lay language so people could connect with the subject on a personal level. This laid the foundation for establishing public policy based on science and facts, documenting the cause and effects, that indicated the importance of preserving the integrity of complex ecosystems. Her work to establish the National Wildlife Refuges and the Endangered Species Act is part of her ongoing legacy.

In their professional communications, women are often cautioned not to be overly emotional or passionate lest it decrease credibility. But Rachel Carson's passion for preserving nature and for the concept that people are part of nature, rather than in charge of it, contributed to her effectiveness. Her passion enhanced by her eloquence gave her a powerful voice creating an image of what might be as well as the potential for harm. Hers is a voice of precaution in a time of heedless indulgence. She set out a challenge that we who know must speak out to defend the natural environment, our only home. She captured the imagination – a spring without bird songs– and offered solutions.

I honor Rachel Carson in my own work by sharing her wisdom and reminding everyone that her voice is now amplified by multitudes calling for the protection of nature in all its complexities and grandeur. She left a challenge for us to take responsibility to speak out from a base of facts mixed with passion and joy. I accept her challenge and hear her guidance in the voices of the trees, the ocean waves, the scent of flowers, and the hum of insects. My aspiration is to live in harmony with nature. I invite you to join this journey and forge your own path in the face of our existential crises.

1.2 ROOTED IN THE EARTH

Rebellion against the patriarchal tradition that wrapped my life like a restrictive corset helped to shape my character. I took relief and sustenance from the struggle against repression when bonding with nature from the earliest days of my youth and it continues to this day. I am a "Baby Boomer" born in 1946, a year after the end of World War II. I was among the children who played under the fallout from open-air testing of nuclear bombs with "bomb shelters" under the house stocked weekly with fresh water and canned food. Mine was the generation of birth control and working in labs without fume hoods and few precautions from exposure to toxins. In other words, I grew up in the environment Rachel Carson wrote about in *Silent Spring*.

It is impossible to know whether the experiences with these multiple and interwoven exposures contributed to my cancer history. No one can say with certainty that any given exposure caused a particular tumor, but I wonder whether my lifetime accumulation of chemicals, radiation, and pollution finally eroded my body's defenses. I am part of the generation born of hope, a child brought into the world in defiance of the hatred and despair of war, crying for renewal and rebirth of the vibrant spirit of human goodness. I grew up in a family dedicated to service through diplomacy and teaching. My aspirations and expectations were rooted in the patriarchal Italian tradition but empowered with women's liberation and the role model of a brilliant and ambitious mother. I became a scientist with a doctorate in genetics who pursued a career in public policy.

The close connection to nature is the greatest gift I received from Pasqualina, my Nona. My earliest memories come from being with her in the garden that fed and nourished our family. She taught me about different life forms and how to notice the details of the individual plants, flowers and insects, both helpful and harmful. I saw the interconnections between everything that lived in the ecosystem of a tiny urban farm, especially the vital part the living system operating in our own lives as a family. When the garden was good, we ate well. When we ate well and thrived, we enjoyed our garden work all the more. This was the foundation of my interest in science.

The courage to bring children into this world and to stand as my own whole person I attribute to Marcella, my mother. A pioneer in the recognition and expression of a woman's power, she taught me to stand for those who were unjustly treated, to stand up for what is right—regardless of the opponent. My birth was her defiance against the scourge and despair of war and the nuclear bombing of Hiroshima and Nagasaki. I was my parents' statement of hope for the future. Their absolute joy from discovering the full dimension of life's glory is something I share with my children. These are the origins of my own reaction both to the ways of the world and to the hardships I have faced. Feeling myself as part of nature from my earliest memories has identified my own purpose in life—to live in harmony with nature.

Rachel Carson's books intersected my life at every important transition. For example, *The Sea Around Us* came to me when I was returning from Brazil to America on an ocean liner and was just entering high school. I stood in the very lowest deck, hanging over the balistrade where I could see the whole of the ocean to the horizon with the boat behind me. I would look into the churning wake of the ocean liner and follow its trail to see a school of flying fish break from the waves. The early time at sunrise was the most breathtaking when the sun first broke over the horizon and cast a golden gleam over the tips of the waves and turned the sky from indigo to pastel tapestries and finally a glowing pale blue. I read in

The Sea Around Us about how the waves were formed, and the importance of the great currents that control the flow of nutrients around the world. I received *Silent Spring* as a graduation present when I entered college. Rachel Carson's concern for chemical contaminants struck home with me because I had been in the Philippines when the apartment complex we lived in was sprayed daily with DDT. As children, we used to run after the spray truck and play in the fumes! I developed a lifelong interest in environmental mutagens.

So many changes have occurred in my life, from television to worldwide communication over the internet; the explosion of suburbs to re-gentrification of formerly industrial city neighborhoods; intercontinental air travel to space travel; consumerism to recycling; from V-Day to Earth Day. But traditions persist in the face of all these changes. The role of women has evolved in my time, and I stood on the rough edge of the transition, torn between wanting to please my family while also soaring to my lofty goals. Frustrated at the preferential treatment and privileges of my brother, and even my much younger little sister, I struggled to maintain an independent path.

I mastered the art of cooking and cleaning while Mother was working, unconsciously supporting her defiance of tradition as she took a full-time teaching job when I was in seventh grade. I became chief cook and housekeeper, architect of the perfect dinner party, and mediator of sibling disputes. But I was not allowed to date, and I had little opportunity for extra-curricular sports, clubs, or peer relationships. My early aspiration to be a doctor was dashed when I failed calculus and realized that a career in medicine would not be mine. Though I finished my doctorate in Biology and Post-doctoral positions at Yale and Boston University, my decision to have children cost me the traditional tenure track of academics. So I found my way to a career in public policy, interfacing with scientists, economists, and lawyers. Rachel Carson had paved my way as a scientist in the public policy arena. She wielded the power of the pen in public, but I became the policy "wordsmith" behind the scenes for much of my career.

Amid all the changes wrought by human ingenuity, my ongoing connection to the living earth remains steadfast. I have been blessed with insight and perception into the wonders of life from Nona who taught me to notice the small miracles of nature as peaches grew from fragrant blossoms to sweet juicy fruits nourished by the straw mulch from a chicken coop. Water, sunlight, and fertile ground created food, as they did in Italy, even in the polluted, dense air that blanketed Pittsburgh in the '50s. Making wholesome food elegantly presented for those I love has always grounded me in the best of my family's foundations.

The other essential ingredient for life to thrive is the caring of community—the love of people connected through the ties of family, culture, or common needs. This circle of caring can be as small as the immediate nuclear family or as large as the whole of humanity. I have always understood my community to include the trees, flowers, and animals whose space I shared. Indeed, there were times in my life when a tree has been my best friend. In those difficult years between childhood and womanhood, I learned many secrets of life sitting high in a mango tree. I will share some of them a bit later.

The decision to have children and raise them to be self-sufficient, independent, and creative individuals defined my first break with tradition and set the scene for everything that followed. Sharing Nature with my children has given me some of my greatest joys. Seeing them grow to appreciate the world around them, deriving inspiration from the beauty of the earth, has been my own reward for bringing these two, wonderful people into this world.

These are the origins of my own response both to the ways of the world and to the hardships I have faced.

In the later times of suffering the indignities of fighting cancer through surgery, chemotherapy, and rehabilitation, I drew on the strengths nurtured in my early days. The foundation was set that allowed me to discover my path as I ran counterpoint to the mainstream.

1.3 FROM HEART TO HARVEST

My earliest memories come from growing up in the Mt. Washington neighborhood of Pittsburgh where we lived upstairs from my grandparents. The backyard was planted end to end with a garden that fed the whole family. We had fruit trees on the borders—pears, plums, cherries, peaches, and of course, fig trees Pop carried here from his hometown of Campolieto-Campobasso in Italy. No space was wasted, and the skill of ages went into making everything flourish. I learned from an early age about planting onions among the tomatoes and putting the zucchini squash and beans in alternating plots each year. The brick walkways had chamomile and thyme growing in the spaces so walking through the garden was a fragrant experience. Mint, oregano flowers, and rosemary scented our linen drawers.

My grandfather worked for the Pennsylvania Railroad as a skilled stone cutter, laying the rail beds that wound through the city and into the boroughs beyond. He went off to work in the morning with a lunch box filled with homemade bread slathered with lard, also containing the bounty of the season—fresh onions and tomatoes with basil in the summer, dry sausage and roasted peppers in the winter, and with all possible combinations thereof. As a child, I did not know we were poor because we ate well. Of course, it was day-and-night labor that fed the family. All summer we preserved food in jars, working under the porch in the shade of the grape arbor that grew up to the second floor. Since every house on our block was farming as we were, we pooled resources to harvest and can. Three or four women chatted away in Italian as they peeled, pared, pressed, and stirred vast vats of sauce over a black coal-fired stove. As children, we went to each cousin's house in turn, learning the seasonings and recipes . . . none written down, just passed on from hand to hand.

At the end of the summer, the cold room in the cellar would be filled with quarts of tomato sauce, ratatouille, beets, carrots, beans dried and stored, onions and garlic hanging from their braided stalks, roasted peppers stored in oil, eggplant and all manner of pickles. In the winter, we would make salchiche (sausage) and Pop would take a good piece to make some prosciutto, a five-year process of packing a pork brisket in salt and pepper. The new one would go to the back of the line, and the front one would be cut. A curl of the deep red meat eaten with a cold slice of melon was a special summer treat.

We did not always have meat, but we kept illegal chickens under the porch that supplied eggs, and vast amounts of chicken cacciatore for the gatherings of the extended family on feast days. (Nona celebrated the saint's days, but not the birthday of her children.) As I have tried to become less dependent on meat in my diet, I have recalled so many of Nona's meals. She made yellow and green zucchini with onions, peppers and tomatoes seasoned with garlic, basil, oregano, and hot pepper to spoon over polenta. This was served fresh sometimes in the summer, with eggs poached in the broth and big slabs of hot bread. In the winter, the memories of summer days would rise from the jar as we opened the jars of ratatouille. And nothing is better than home canned tomato sauce over hand-made gnocchi or pasta.

I have always put up in jars some freshness of summer for winter enjoyment. I do this partly because I love jam on my breakfast toast but dislike the overly sugared and tasteless commercial preserves. I find that when done in small batches of about 12 cups at a time, it is simple and very rewarding to make jam. Strawberries in May, rhubarb, plums, peaches, raspberries, and blackberries all find their way to the larder. It is also easy to put up fresh fruit, especially peaches and pears and applesauce with just a boiling water bath as a processing requirement. And of course, the fresh tomato sauce, ratatouille, beets, carrots, and eggplant come in season. The time is easily found when the reward is so tangible and adds so much to the quality of life. Even though I rely on my Community Supported Agriculture share from Kretschmann Organic Family Farm for most of my produce to can, I still feel connected to the farming tradition of my family. It warms my heart to share Nona's legacy with those I love.

The modern way of relying on processed and prepared foods never made sense to me. I retained the tradition of lovingly making good wholesome food by hand for my family. Feeding people to provide comfort as well as a sustenance has become a personal counterpoint to the homogenized, sanitized, tasteless, plastic-bound, and commercially prepared fast food. Now more people are concerned about contamination and want to live less intensively on the earth, I see the art of home cooking and canning as skills of resilience for the future.

One summer, after a hike with my son Steve in the Emerald View Park in the Mount Washington neighborhood where I grew up, we stopped by the house where my grandparents had lived. The same privet hedge that shielded the front porch from the street was there with a swinging gate, but the flower bed of sweet William and zinnias Nona always had planted was no longer present. The lady who lived there had bought the property from my grandfather and let us go down to the garden in the back. She prattled on about how they had to take down the grape arbor to add an extension to the porch and had taken years to "get rid of that fig tree" in the middle of the yard because the fruit attracted wasps. I looked over the sun-scorched expanse of "lawn" on packed earth, with a car parked where the tomatoes had once stood high and heavy with fruit. Against the fence at the border with the neighbor, I spied the remnants of a grape vine struggling along the edge of the mowed lawn.

In that moment, I realized the accomplishment of Nona and Pop in converting this heavy shale ground, covered with rough grasses, dusted daily with coal soot and sprinkled with acid rain, to a fertile farm plot that fed our family for fifty years. In contrast to what was touted as the American Dream—a grass yard for kids and dogs to run around in with maybe one tree—my grandparents gave their sweat, skill, and love of the land to break up that ground and enrich it with manure so we could eat better than the wealthiest people in Pittsburgh.

Laying Down Thread

I learned embroidery from Nona when I was six or seven years old. She taught me to sew by hand turning the heavy cotton flour sacks into dish towels or aprons. The 25-pound sacks were emptied into the flour hopper in the kitchen, with a sifter at the bottom to sift flour into a bowl when needed. In Nona's house, a large batch of bread appeared once a week to serve the working men of the house with lunches, and the rest of us with nourishment. The empty sacks were washed and taken apart to lie flat, and I learned to draw a thread for a straight seam, and hand-turn a hem.

Nothing was sent to use without embroidered embellishment—at least a prayer or blessing in white thread on white cloth, or more elaborate freehand flowers trailing along the edge. Counted thread work and smocking decorated aprons. And of course, the darning of socks, turning of shirt collars and cuffs, and mending were endless tasks. These simple chores occurred in the evening after dinner and were embellished with rich conversations. Nona was the designated point of contact between the community in America and those in Campolieto Campobasso because she would help her friends read their letters in Italian and translate them, as well as help to write responses in Italian for the family and friends back home. During this process, gossip flowed freely, and commentary on the fortunes and politics in both countries punctuated the conversations.

In one of my earliest memories, I recall a Saturday summer afternoon on Nona's back porch. The grape vines were so thick that the heavy leaves shaded us from the sun. I sat on a little stool at Nona's foot with my embroidery hoop working on a set of pillowcases to be embroidered with flowers spilling from a basket. Mrs. Nichola, Comare D'Alessandro, Aunt Bernice, and Aunt Matilda were there, each with some hand work, were all talking and laughing as they shared stories with a mixture of Italian (mostly) and some English.

I was oblivious to the content of the conversation but remember that I deeply connected to these ladies who could laugh and be happy despite the hardships and separation from their families. I would show Nona my work, and sometimes she would be pleased and give me the next color to add to the design, or sometimes she would tell me to take it out and start over. The back had to be pretty, too. No tangled messes were tolerated. Sometimes she would send me to look at a flower growing to see the shape and the detail of leaf and flower form. The embroidery translated the mental image to the cloth.

As I grew older, I came to treasure these Saturdays with Nona, mending, embroidering, and sharing time. We talked of problems and fears and hopes. I marveled that she had so much wisdom and strength. Bare-root grape vines and fig trees came to America in Pop's pockets now grew to shade the second-floor porch and offer fruits to eat with cheese and bread and wine as we talked. Nona listened to my struggles for independence from Father's rules and from the unfairness of a woman's place in the world. She was wise in many ways. She told me "The men may rule, but the women govern." I watched at the family gatherings over Sunday dinner where all major decisions were made. Pop would declare the outcome, but Nona guided the discussion and arguments, sometimes with force, but most often with a well-placed question or observation. And it was Nona who executed the logistics and details.

Nona became my refuge as I struggled to become my own self within the constraints of the family. I was a child of the transition from the patriarchal ways of the Old Country to the modern ways of liberated American women. Mother struggled for her place as a working woman with a career teaching in the public high school, and as something more than a diplomat's wife when we were overseas. But the family strictures on my personal freedom remained tight. Nona shared her insight about managing my independence. She shared her stories of defiance against her own family boundaries, marrying for love instead of the chosen person who would make a more prosperous match, according to the traditions of her family in Italy. When Pop had saved enough money for passage and had a home for them in Pittsburgh, she came to America with her five-year old son, my father, and her two-year old daughter Rosa, never to see her parents or family again.

She helped me see the poet's heart under my father's bluster and strictness. The boy she sent off to war had come home deeply wounded emotionally and closed off the song in his heart. When we sat together doing the menial chores of mending and bringing beauty to mundane tasks, I learned the value of patience and building up the strengths of other people. Her ability to see an achieved goal in the future while not fearing the trials that stood in the way gave me courage. Her faith rested in prayer, for she said the rosary in Italian every evening, and walked up Lelia Street to St. Justin's Church at the top of the hill every morning for 6:30 a.m. Mass. My faith grew from seeing the cycles of Nature and finding beauty in the world, no matter what evil or hardship life presented. She cultivated the conviction that if you hold love and beauty in your heart, you will have strength to fight whatever battles you face.

I think about the many embroideries I have done over the course of my life. They were mostly given away as presents to other people or sewn on children's clothes abandoned or outgrown as time went on. Today when I embroider flowers whose forms manifest from mental images through thread applied to an image one strand at a time, I think about Nona and all she endured to make her family a better life. Embroidery has a way of capturing one's sadness, loneliness, or fear and making it beautiful. I lay down thread and remember my brave Nona.

Castle Shannon Roots

When I was five years old, my family moved from the upstairs of Nona and Pop's house on Southern Avenue in Mt. Washington into an old farmhouse in the Castle Shannon section of Pittsburgh. It was a duplex, and we were going to live in the bottom part and rent out the upper part. My father worked for Buhl Planetarium in those days. On Saturdays, he would take me with him to work, and I would sit on his lap and watch what was called the Sky Show. He was the narrator of the story of the skies and the constellations in a big theater with a domed ceiling. People would go in and sit down where it was totally dark, so dark that they could not see their own hands in front of their faces, and so quiet, we could hear ourselves breathing. It was scary and delicious, because I could snuggle against my father and feel safe and warm. Then would come the rumble of his voice from deep in his chest and he would tell the story of the stars.

"Long ago and far away the Milky Way was formed . . ." As he talked, there rose from the pit in the middle of the room a great Zeiss Planetarium Projector with many cameras mounted and directed to the ceiling. It rose up casting the images of the universe onto the dome. That projector looked like a huge, many-headed monster as it rose over our heads, a fearsome mysterious creature that spewed stars. I would sit there and listen to the legends of the constellations and think about how the sky looked at night.

When the show was over, I was allowed to be a

helper in the static electricity show. One of the scientists in a white lab coat, wearing horn-rimmed glasses along with a hairless head would talk about electricity. There was a Van der Graaff Generator there, and when he turned it on, big sparks flew out of it. They were blue and purple and made a cracking sound like lightning with a buzzing sound in between. The scientist would hold a wand, and would hand it to me. When I held it near the Van de Graaff Generator, the hair on my head would stand on end like a cloud around my face. It tingled and I would walk around the circle so all the other children could see the effects of static electricity! Then the scientist would explain about static electricity and how we could rub a comb, and have it pick up particles of paper, or rub a cat's fur and create sparks when we touched things. I enjoyed being part of the scientist's exhibit and seeing other children delight in gaining knowledge.

Usually, we were on our own to play and find amusement when the adults were working on the house. An old lilac tree stood near the kitchen door of the house. It was so big that there was a hollow place near the bottom, and a curvy branch that came out to make a nice little seat. This was a particularly interesting place because there was a chipmunk hole there. We would see the chipmunk running in and out, and if we sat very still on the edge of the kitchen steps, he would come close enough for us to see him very well. So, we started leaving him bits of peanut butter sandwich on the curvy branch. Sometimes he would come right near us to eat the peanut butter, but if we even breathed too loud, he would be gone in a flash back into his hole.

We had a big garden at this house. Father and his younger brother, Uncle Al, my grandfather Pop and Nona's brother, Uncle Louie, worked to clear out the brambles and brush. They raked and smoothed out all the roots and mess and made a garden. Pop let us help him plant. He had a pitsuke, a short-handled shovel that was rounded on one end. He would poke it into the ground and make a hole, and we would drop a seed in and cover it over with soil. Then we would water the row. Some of the plants produced hot Italian peppers. When the plants grew up to bush size and had ripe peppers, my brother Mikie would go to inspect them with Uncle Louie. One day, Uncle Louie taught Mikie to pick the pepper and eat it right there. I thought it would be too hot to eat but Mikie learned to like hot peppers picked fresh from the garden with nothing else! I did not like hot peppers, still don't because once when I tried one, I got some of the juice in my eye, and it burned for a long time.

This garden was twice the size of the one at Nona's house, so we planted lots of vegetables to share with the whole family. The various kinds of tomatoes grew tall here because there was good ground and lots of sun. I especially liked the ones that looked pink, with seeds scattered throughout the inside, instead of being in jelly cells. These were beefsteak tomatoes, and they were delicious. We also learned how to help weed the garden.

One time, we were helping, and trying to take out the weeds in a row of beets. The plants were still very small, and we were not really sure what the beets looked like as young plants. So, we accidentally pulled out the beets but left the dandelions. Uncle Louie was quite upset. He yelled at us in Italian and we ran away. Later he was sorry we were frightened, and he gave us each a shopping bag. He promised we would earn a quarter each when we filled the shopping bags with dandelion flowers—just the flowers, no stems or leaves. Of course, we thought this was a great idea, so we ran out to the front lawn and started pulling off the flower heads of all the dandelions.

When we finished our own yard, each of us did not have a full bag yet, so we went to the neighbor's yard next door. This place was owned by two elderly ladies, the Starrett sisters, where we pulled the heads off all their dandelions. When they saw what we were doing in their yard, they were very curious, and called us to the kitchen for cookies and milk. We told them we were collecting dandelions for Uncle Louie. We picked all afternoon, even along the edge of the street, and in the playground at the end of the block. Finally, when we had enough to fill both shopping bags, we took them to Uncle Louie, who inspected them and gave us each a quarter.

He took them down in the cellar and dumped them into a big tub. He washed them with cold water, then covered them with water and added sugar and put a lid on the tub. Every day, he would go down and stir and tend the dandelion mess. It smelled really awful. Mother complained bitterly about his concoction, but he just smiled. Soon, the dandelions began to ferment and foam. And little by little the fermentation turned into alcohol. Uncle Louie pressed out the remains of the flowers and collected the liquid. I am not sure what happened in the next stages, but later that year we saw several bottles of clear golden wine that smelled faintly like dandelion flowers. When anybody got sick, we had a brew of chamomile tea steeped with a dried fig, honey, and lemon. This would be strained into a cup, and a generous dollop of dandelion wine would be added, which was guaranteed to make you sleep. Uncle Louie often drank some for his "gout."

Sometimes we would go with him on the streetcar back to Nona and Pop's house in Mt. Washington. We would get on the Castle Shannon streetcar and ride into the junction in Mt. Washington. There, we could either take the #40 streetcar that went down to Southern Avenue, or we could walk up the stairs and down Lelia Street. Sometimes the streetcar would be crowded with people. That was when it was good to be with Uncle Louie. He used to eat raw garlic for breakfast, claiming it made his blood strong. I still doubt that claim, but it certainly made his breath strong, especially when adding daily doses of dandelion wine.

So, when we had a crowded streetcar, he would

stand over some unsuspecting person, hold onto the overhead strap, with us each holding onto the loops in his dungarees, and then he would lean over and say "Nice-a Day, HAAA!" If the person did not faint immediately, he would ask them how they were doing, and where they were going. Usually, they got up and moved to get away from the fumes. Then he would sit down with one of us on each knee.

There was a grape arbor in the back yard that was tall enough for us to play under in the shade. My Father had rigged up a swing, and it was cool even in the summer. We would go there to play or just sit and watch things in the garden. In the autumn, the grapes hung heavy and sweet so we could pick them to eat right off the vines. This was a great pretend place where we made up games and devised all kinds of projects. We would take the smooth white stones out of the gravel pile Pop had for paving the cellar floor or putting in a driveway. We would only take the smooth white ones and build castles and parapets by making excavations among the roots of the grape vines. We had flowers to decorate with, and we made up all kinds of tales for our imaginary protagonists.

We also made a "recreation park" for the Japanese beetles. We were required every day to pluck them off the roses and vegetables growing in the gardens, so we used to find ways to play with them. We would tie a thread around their head or thorax and have contests to see which one would fly the farthest. We also, rather cruelly I'm afraid, pulled off their wings and made them swim in our palace pools. They could not fly away, so they would paddle in the water while we pretended they were visitors and that they liked it. Their final destination was a jar of oil so they would no longer eat the garden plants or the roses. This form of pest control would have met Rachel Carson's approval, though she would have been horrified at our games, I fear.

One day when we were outside playing under the

grape arbor, Mom came out to tell us that soon we were going to be moving away to South America, specifically to Bogota, Colombia. From that time, my life took an entirely new path. We would travel all over the world, but Pittsburgh was home, and we would always come back in between foreign service posts. My own roots ran deep in those gardens of Southern Avenue and Castle Shannon. The bonds formed with the land there have sustained me all the days of my life.

Rachel Carson wrote about the sense of wonder that rises innately in children when they are given the freedom to explore the natural world. There is no more fertile ground for developing imagination and self-sufficiency in children than to have time to fill with playmates, unfettered by structured activities. To lie in the grass and imagine shapes in the clouds, or to sit in a hammock in a lazy summer afternoon listening to the bees hum and the birds call their chip notes instill a permanent bond with nature.

1.4 BLUE BUTTERFLY

Father joined the U.S. Foreign Service working for the United States Information Service, now part of the CIA. As children, we never really understood what he did for his work, and questions were discouraged, but we had the opportunity to travel the world with him. My Father was stationed in Rio de Janeiro Brazil from 1956 to 1959. My parents did not want to live in the high-rise apartments offered by the embassy for diplomatic staff because they saw no space for children there, and it was in the middle of the dirty, crowded, and sometimes dangerous city.

After several weeks of temporary residence in the Gloria Hotel, my parents found a house to rent built into the cliff above a small beach around the curve from Copacabana and tucked into a bay before Ipanema. It was a small, obscure place, once dominated by a convent out on the spit of land occupied by a small village of fishermen. Above us, a mountain loomed with a huge hole created from the fall of a large pieces of rock that now rested both in the front of the house and in the bay below. This terraced house on three levels came with the services of Manuel, the gardener, who served as the representative of the owner. He was our guide to the wonders of the jungle that covered the mountain behind the house.

Mother immediately took over the second terrace behind the house to make a vegetable garden in raised beds. She sent for seeds from the catalogues and grew lettuce, cucumbers, beans, and tomatoes. The kitchen was open to the outside, so the shelves Father built to hold the food had each riser set in a pan of oil to keep bugs from crawling up. Of course, if anything on the shelf was pushed against the wall, the entire shelf would soon be overtaken with ants. Father built an extension connecting the very small kitchen space with the laundry shed, which was also living quarters for the maid. A temperamental pump brought water once a week by pipe up to a tank on the roof, which also collected rainwater. The gravity-driven water was brown, and often had things swimming in it. So for drinking water, this was boiled and dripped through a porcelain filter, able to process five gallons every 24 hours. This we used for cooking, drinking, and brushing teeth. All other water, including for bathing, was boiled then filtered through a linen cloth, which was changed daily. We had a chicken coop with a dozen chickens and a rooster behind the guest house on the third terrace. This was our playhouse where we kept our toys and our launching point for jungle expeditions.

Jungle Adventures

Our parents forbade us to go there alone, because beyond the next house, was the *favela*, the slum that overlooked Rio and was populated with people living on the edges of desperation. The children would watch and rapidly plunder any garbage we threw out. I am amazed that we produced such a huge amount of garbage buried in the pit at the top of the property. The plastic, the cans, and the detritus of American stuff was collected in the night and put to use. Our early forays into the jungle were only with Manuel. He showed us how to walk with a stick in front of us to cut through any spider webs that crossed the path. There were banana spiders as big as my spread hand with webs that would stretch across the entire path. We learned to be sure the hanging vines were attached to something sturdy before we would swing on them; we learned not to sit on the mounds of termites that looked like inviting stools; and we learned to watch out for snakes.

There were so many new and strange things living everywhere. The potential dangers receded from consciousness as we became more familiar with the ways of the jungle. Escaping from Manuel became our favorite game. He soon figured out that we were fine and took a nap while he was "watching" us. The jungle was an amazing place, with a canopy so thick the shafts of light would filter through like the fingers of God. The pink and magenta bromeliads grew everywhere, and small yellow and brown orchids cascaded from rotting logs along the path. There, the butterflies would catch the light as they fluttered around in search of nectar. The brown undersides with iridescent blue upper sides gave perfect camouflage when at rest among brown leaves or bark. Some had clear patches in their wings that gave simulations of light flickering though the leaves.

I became fascinated with these spectacular creatures and began to watch and study them. I sought out their egg masses and made daily visits to see the caterpillars eating leaves, then forming a chrysalis, and emerging as adults once more. I never actually caught one enclosing, but I saw caterpillars molt and I watched one form a chrysalis. I was fascinated with the transformation that took place in that little resting pupa, and I asked questions of my teacher, Sister Modesta of Our Lady of Fatima, where we went to school. She assured me that it was a miracle of nature wrought by the Holy Spirit but gave no real explanation.

During these years when most pre-teens were bonding with their peers and forming relationships outside the family, I was isolated by both location and the oddness of my interests. Because we did not live in the American compound, we did not have neighbor children to be friends with. The English family that lived close to us had two little girls, but they were much younger, and had a nanny who kept them close to home. My classmates thought I was distinctly weird, always in a book, and too shy to speak to—who also had a strange accent. As part of the diplomatic service, we did not have a military rank to define where we stood, so we were designated as inferior creatures to be bullied, teased, and shunned by the ranked military children. I did not fare well in this system.

I withdrew by reading Nancy Drew mysteries and sitting in trees so I could be alone.

For my eleventh birthday, I received a Fisher Price toy microscope with the ability to magnify up to twenty times and a mirror to reflect light from below the subject. I literally fell into the microscope that year. I looked at the butterfly eggs on leaf undersides and examined them to see the geometric shapes and the patterned array, unique to each butterfly species. I learned to make wet mount slides and was fascinated to see the cell patterns as I peeled leaf layers apart. Nothing was beyond examining—moss, fungi, insect parts pulled from dead beetles found on the ground, and my brother's hair.

The mango tree that overshadowed our play-house soon became my personal refuge. The branches were so wide that I could put a pillow down to sit on and lean easily against the trunk with my legs stretched out in front of me. I could see the ocean below and feel the breeze that wafted up the mountain, cooled by the layers of vegetation as the currents rose. I learned to sit very still and melt into the tree, so the birds would come close. The warm scented jungle air would caress me, and I felt the life force pulse through me from the tree against my back.

Sometimes when I was sitting there alone, I would hear the small tree monkeys screaming to each other. I would pretend in my head that I was one of them and had their agility to leap from branch to branch and scamper everywhere in a group. When I sat in stillness, hiding, feeling sorry for myself, and barely breathing, the little monkeys would come close enough to eat pieces of banana speared on the twigs close at hand. They were very mischievous creatures. Once they entered the bedroom of the very exotic lady who lived next door to us, on the other side of a tall brick wall. They came out trailing brightly colored scarfs which they brandished like trophies as they fled the shouted curses of the maid who chased them away. They made me laugh. The beautiful blue butterflies would flitter up from the forest floor on the draft of the warm afternoon exhalation of the jungle. A fat caterpillar creeping along to find a safe place for its chrysalis showed the determination and confidence in the metamorphosis to come. I wished I could go to sleep and wake up transformed into a beautiful butterfly, with that caterpillar's confidence of success.

The Quartz Beach

The beach right below the house defined the soundscape with the rhythm of the waves, sometimes a sibilant murmur in calm times when the waves rolled in and flipped over like a row of teacups. Other times, the waves were higher than the house-sized boulders and crashed wildly, making echoes against the rock face of the mountain. We would go down early in the morning to watch the fishermen bring in the nets, and if we helped pull them in, they would let us take a fish home. I asked for a flying fish, and they laughed because those are not edible. There was a huge skate pulled in, and everybody was jumping to stay out of the way of its stinging tail. And once an octopus was pulled up, but it escaped into the sea when a wave rolled over an open part of the net. It swam into the water with its graceful pulsing motion and disappeared into the deep.

We loved to explore the quartz caves along the edge of the beach at the rocky end. Usually our parents would be resting on the sand, or my father would be perched on a rock, waiting to photograph the perfect wave. One afternoon, I walked with my brother and sister along the sandy edge around the bend of the rocks, and looked into a cave. The sand was smooth and soft, and there were interesting ripples in it from where the water had been. We were exploring and pretending we were pirates hiding treasure, when I looked back toward the entrance and saw that it was half obscured by water. The tide was coming in! We would have to swim under to get out. So, we immediately went to the entry, and told Linda to take a deep, deep breath and hold it. Then we all dove under, holding hands, and came out into the sun. Three waves later, the entrance would be completely covered with water. We never told our parents about this, or our beach privileges would have been canceled forever. We had to go back by climbing over the rocks because the beach of sand we walked in on had been inundated by the incoming tide.

Fruits and Flowers

The land our rented house sat on had been an experimental plantation once where coffee trees were planted in the shade of banana plants, yielding two cash crops. It apparently did not do well, but we were the beneficiaries of an abundance of both coffee berries and bananas. There was also a variety of fruit trees on the property: guava, papaya, mango, limone (a sort of sweet lemon hybrid with grapefruit), and a row of orange trees that grew on the patio of the first level with their tops just reaching the patio of the main floor.

Of course, the first thing Father did was to prune all the fruit trees, with much hand wringing and desperate protests from Manuel. Within a few months, new growth, blossoms, and then fruit came in abundance. There was an espaliered fig trellis growing up the wall to the house next door that gave super sweet figs the size of plums. I have never enjoyed so much amazing fruit, and the bananas were especially delightful, unlike the pale almost tasteless versions we knew in the States. Surely the abundance of sun and ripening on the plant made a totally different taste and texture. My small bedroom faced the ocean, and I could feel the rhythm off the waves as I would fall asleep. The fragrance of the orange blossoms constantly infused the night air, a distinct contrast from the sulfurous fumes of the coal furnaces in Pittsburgh.

The flowers were amazing and ubiquitous. The vines of yellow trumpet flowers grew up to the main patio and rimmed the fences with shiny leaves and bright flowers. Linda and I would use them to make ball gowns for our dollies. Orchids were so profuse, that we had them in the house all the time—brilliant magenta and brown and yellow and white varieties. The bright pink bromeliads sprouted everywhere out of the tree bark. When I think about my time I Brazil, I immediately call to mind the fragrances of the flowers and the delicate butterflies that flitted among them.

My father had fenced in a small shed ringed with banana plants to serve as a chicken coop. My mother bought six young living chickens and a rooster from the market to inhabit the space. We had the chore of checking the egg boxes and taking the eggs in the morning after spreading crushed corn and our despised leftover oatmeal for the chickens. One morning we went up to feed the chickens, and out came one of the red hens with a brood of ten chicks in her wake. She had made a nest up in the clump of banana plants and hatched out a brood. They were adorable and we were so excited to have them. We watched them get bigger and feather out. My mother clipped the wing feathers so they wouldn't fly away, and of the ten chicks, four were roosters. She gave one away to a neighbor, and we ate the other three. I remembered that Nona had prepared chickens for eating, but Mother banished us and Manuel did the butchering when we were at school.

Special Baskets

There was no modern grocery market in Rio when we lived there, though an Acme Market opened just before we left. So, we obtained supplies from the post exchange, which was fine for frozen meat or canned and packaged goods, and canned or powdered milk. But for fresh produce, Mother went to the open-air street market that came to our neighborhood on Wednesdays. She would go with the maid or Manuel for assurance and advice on the pricing, and had to bargain in Portuguese to make purchases. She had a set of specially shaped baskets.

There was a brightly colored covered basket for eggs or small fruits, large open baskets for carrying vegetables and melons, and two large, covered baskets for buying chickens to eat (not the ones we raised to give us eggs.) She only bought living chickens because the already butchered meats hung in the sun and were cut to order, usually with abundant flies and other insects crawling all over them. Most of the Brazilians charcoaled their meat and the outer exposed areas were well charred.

Father built a Churascoria (bbq pit) on the second level back area behind the house, and several times hosted large gatherings where the meat was speared on long pokers stuck into the coals until done then served over huge mounds of rice with vegetables wrapped in banana leaves. When my parents had parties, the people from the *favella* would gather on the hill overlooking our house and watch. When the guests left, they always came down to pillage for leftovers or anything left out. Manuel usually kept guard on these occasions.

The three years I spent in Brazil covered my transition from childhood to teen. It was not the usual time of girlfriends and social connections outside the family. The forced intensity of the family shaped my personality, but also gave me independence and resilience. This also parallels Rachel Carson's childhood. She was isolated by space and her very protective Mother from having a close circle of friends while growing up. One of my favorite pictures of her depicts her at about eight years old sitting under a flowering apple tree reading to her dog. I truly resonate with that image. When I am lonely or feel isolated or unhappy, walking outdoors in Nature never fails to restore my sense of belonging. The cathedral of trees became my touchstone. I learned to be confident alone, never bored with the living world around me. We returned to Pittsburgh as I entered high school, but I carried the lessons of the sea and the jungle with me. I had found my center.

1.5 WHO OWNS THIS LAND?

I had spent three years after college finishing my doctorate in biology at the University of Pittsburgh, staying behind when my husband went to Waterbury, Connecticut for his internship after finishing medical school. It was an unusual decision for me to stay behind, and I had to move back into my parents' house. During this time of intense effort, I often stayed in the lab, timing my hours of activity with the cycles of the *Drosophila* larvae I was studying. My fellow doctor's wives were horrified that I was not supporting my husband, and my family was shocked that I was two years married and had no child yet. My auntie was convinced I had an abortion because we moved our wedding date from August to April. I had taken my fate into my own hands and took birth control pills for three years while I finished my degree.

This period of being a graduate student working in close association with a few outstanding scholars grounded my path of independent thinking. Under the guidance of Harry Corwin, my major professor, I gradually learned the skills of critical thinking and honed the craft of giving interesting, well-organized lectures. However, my research interests were unconventional. I had friends who were canoeists and collected Mason jars of effluent from the pipes from which mill water dripped into the Monongahela River. They asked me to run Ames tests on it to see if there were mutagens in the water. Some of the samples had to be diluted 100 to one to let half of the flies live. There were many samples that had mutagenic activity, but the river water itself was not.

I wanted to pursue this inquiry for my doctorate since environmental mutagens were clearly an issue. However, I was advised that if I wanted to finish my degree in a timely manner and join my husband in Connecticut,

I needed to take a topic that had a well-established literature base. If I entered this field of inquiry about environmental mutagens, there was little published research in place, and I would need a decade to establish a base of data on which to rest my work. Besides, there was no way to run a control, with all of the materials in the river being of undetermined origin, since the pipes ran long distances from the source. I would have to begin with analysis of the components, and since the effluent was variable from time to time, and the sources were impossible to identify, there was no clear way to set up an experiment.

Clearly, Rachel Carson's warnings about taking precaution for what is put into the environment held no sway over the rampant industrial engine of the 1960's. I joined the environmental objectors and demonstrators of the first Earth Day in April 1970. I was active in the campus teach-ins and the community outreach to union halls and schools urging the legal protection for clean air and clean water as a citizen right. My outrage at the toxic pollution remains to this day, an undercurrent of disappointment and frustration.

I left Pittsburgh in 1971 after I finished my degree, before the Safe Drinking Water Act finally passed in 1974, and the steel mills were required to control their emissions. When I came home to visit my family, my uncles would regale me with vituperative remarks that my environmental activist friends and I had shut down the steel mills. I encountered this same simmering fury many years later when I was at the Rachel Carson Homestead in Springdale, PA under the plume of the Cheswick power station, still burning coal, and only recently shuttered.

When I was a new mother, we lived in Boston, Massachusetts, then moved to Waterbury, Connecticut. While still in Boston, we would visit my husband's Smith family home in Narragansett Bay, Rhode Island on weekends. It was a relatively short drive down from Boston and offered an escape from the noise and pollution of the city.

It was a chance for the Smiths to play tennis together in good weather and visit with the newest family member, baby Jennifer Claudine. I was still a bit of a hippie in those early days of 1974. I was not a good enough tennis player to compete with the lifelong tennis players of the family as they took the afternoon to play whole sets. My mother-in-law, Margaret, really needed to nap in the afternoon and assumed I would rest with the baby as well.

But I enjoyed the time to take long walks along the shore with Jenny in her baby sling across my chest. I went out in a fringed leather jacket and matching moccasin boots and bell bottom jeans with a floppy silk-flower-bedecked hat to keep my long hair from blowing around and getting full of salt spray. I slung the requisite baby stuff in a flowered hobo sack over my shoulder and we set off one late April afternoon to walk down to the lighthouse that looked pretty close. I strolled down the street to the shore and turned toward the lighthouse wandering along the sand.

Any fishy and marshy salty smell unique to the shore still evokes memories of those days and many others spent on the seashore. We stopped from time to time to watch the antics of the gulls as they wheeled and dove into the waves. We visited with some little boys digging quahogs and threw driftwood sticks for their puppy to retrieve. At one point, we sat on a rock and had animal crackers and juice, and I cut up an apple. Jenny was delighted to throw pieces of crackers to the hungry gulls that soon thronged our perch. We watched a sailboat tack around the bay with a billowing white sail and a whole stream of signal flags. There were sea grasses and some waxy flowering plants along the shore in some places to add texture and interest to the scene. I was wishing I had brought a sketchbook to draw them and made a note to add that to my walking gear. I also lacked binoculars to look at the shore birds up close. Was that an osprey nest on the pole across the bay? It was impossible to tell.

As we continued walking around the point, Jenny fell asleep and rested softly against me. I was feeling connected to her and to the rhythm of the waves as they rolled into the shore. As I walked along, the shore became more rocky and steeper, and I finally noticed that the band of sand was getting much narrower. I realized that the tide was coming in. The rocks were covered with barnacles, snails, and seaweed, so I knew this beach would soon be under water. It was too dangerous to try to walk on the stones with the baby challenging my balance. And though I would not hesitate later in the warmer summer to walk along in the incoming waves, it was still too cold to be wading in moccasins. I did not see a public access pathway to the road from the beach, so I looked for a place that was a little less steep. As I clambered up the rocks onto a broad expanse of lawn, I saw a driveway that would take me to the street and headed in the right direction.

Before I had gone twenty feet, I was accosted by two large dogs barking at me and bounding around my legs. Jenny woke up and started to cry. I froze in place and called for help. An angry-looking older lady came out of the house calling out that I was trespassing on private property and she had called the police. I begged her to call off the dogs as I tried to explain that I was unfamiliar with the area and was caught by the rising tide. She did not believe me and kept stating that a rash of vandals and petty thieves was active in the area. She clearly considered me a suspicious character, so she was being cautious. I guess I didn't really look like a doctor's wife and mother as I stood there on the verge of tears.

I tried to comfort Jenny as I continued to explain that I was visiting my in-laws and asked if she perhaps knew Margaret and Waldo. By now we had made our way to the front of the property, and I was in sight of the public street and sidewalk that would allow me to go back without trespassing. She finally relented in letting me go just as the police drove up. I had to explain the whole story all over again, and the woman decided not to press charges this time. They let me go with a warning not to trespass on private property again. The police car followed me for a few blocks from Briarcliff Avenue to Warwick Neck Avenue until I turned down Narragansett Bay Avenue, and they turned the opposite direction and went on their way.

As I approached the Smith house, I encountered Mrs. Slater, a member of the Rhode Island Senate, who was weeding her garden next door. I stopped and talked with her as she cooed over Jenny. I related my adventure and asked her why there are no public footpaths connecting the shore to the public streets. She explained that people really do not want "the public" at the shore in front of their private houses. She said that there were a couple of public access areas that go right down to the water, but they were really designed for boat launches, not pedestrians. I pointed out that below high tide lines, the shore was technically public land. The whole thing seemed quite unfair to me, as if anyone could actually own the sea!

I am amazed at the arrogance of the belief that we "own" land, especially on the seashore. I reflected on the long history of human habitation in this space. The Narragansett tribe lived and fished there centuries before and their names still marked this area—the Puttaquomscut River, Pawcatuck River, Winnesquam Drive, Chepiwanoxet Island, and so many others. The resilient ecosystems of the bay with its marshes, reefs, and rocky shoals persisted where they were left in a natural state, even when interrupted by man-made docks, walls, and marinas. The life of the edge of the sea persists despite the intrusion of humans.

I felt disconnected from the people who have the privilege of living along the bay but tried to control and contain it with sea walls and rocky borders. Their access to the country club mattered more to most of them than access to the edge of the water. For me, being able to walk unimpeded along the shore is a priceless gift that no person can own. The class disparities in control of land were etched into my mind on this occasion, and I reflected on this issue as an undercurrent to so many of society's problems. The words of Chief Tecumseh resounded in my mind: "The land is given for all for the use of each with love and respect."[3]

1.6 WATER STUDIES

Ever since I spent three years in Rio de Janeiro, I have felt a connection to the ocean. Later when living in Connecticut for 23 years, I had easy access to the shore and went there often, sometimes with the children and their friends, sometimes alone on my lunch break.

I would go to Hammonasset Beach State Park on Long Island Sound and dance on the sand of the water's edge, moving back and forth with the advance and fall of the waves while trying to feel like water. This beach was edged with stones and large rock formations at Meig's Point that challenged our dog Bootsie but enthralled the children. They clambered over the rocks to stand at the heights like mountain goats. I let them explore as they wished, but kept an ear out for distress, or unusually long silences. I was an embarrassment to my children at the beach, not only by dancing in public but because I came swathed in a long, red kaftan, knotted at the hem to keep it out of the water, and an enormous straw hat and scarf to keep the sun off me as much as possible.

Instead of stretching out on a beach towel under an umbrella with a novel like other proper mothers, there I would be, much like Rachel Carson, squatting over a tide pool, sometimes with a magnifying glass and watching the antics of the hermit crabs or seeing what was under the stone ledges. Small crabs, snails, and several mollusks populated these shores. I was tempted to harvest the blue mussels as they clung in massive arrays over the rocks interspersed with barnacles. However, I dared not gather any to eat because the water of Long Island Sound was not always safe and clean. My children, Jenn and Steve, would come to see the hermit crabs and tried to get them to race each other with limited success. I would crush a snail so they could watch the crabs come to eat the pieces.

With the help of a magnifying glass, we could see barnacles that were under the water line pushing feathery extensions into the water to sweep up nutrients. We would find the egg cases of sea snails at the tide line. I opened one to show the embryos inside, curled together until they would emerge as tiny, shelled creatures to find their way into the great expanse of the sea. We would walk along the high tide line and pick up shells and interesting stones, sometimes finding the cast-off flotsam of civilization among the driftwood and seaweed.

Once in a while, there would be strange and frightening items there, like needles and syringes, probably from debris of insulin users cast out in the trash. Long Island had a trash dump, and often items would blow into the water from the garbage scows that served it. There was some plastic debris, beach litter carelessly left by other visitors, which we were diligent to pick up what we saw—harbingers thirty years ago of the plastic inundation that now covers shorelines everywhere.

The salt marshes on the edge of the beach offered a wonderful wetland habitat to explore. Steve attended a summer camp there and brought home a preserved razor clam he had collected. I asked him to tell me about it, and he reported through tears that the poor clam had put its footie so far out trying to get away from the preservative, and that he felt sad to take it away from its home. I thought about Rachel Carson then, who studied sea creatures and measured and recorded their characteristics, but always put them back in the water when she was finished. Steve has an empathetic and sensitive nature that was not destined for the analytic world of science or engineering. The music and rhythms of life have always lived in him, and now music is his medium of expression.

We had a tradition of going to the shore on New Year's Day. We would bundle up and wear waterproof boots and set off to Hammonasset. The winter beach has its own beauty in the form of the waves and the patterns on the sand in the play of light from the pale winter sun. The scene was stark, with the cries of the gulls piercing the sky as they wheeled overhead. The waves made ice patterns as the tide receded, resembling fairy castles that sparkled in the slanted sunlight. The wind used the grasses along the shore to create patterns in the sand. Sometimes we could follow little tracks from creatures hiding in holes under the dunes. Sometimes the snow would be piled there too, adding contrast with cover of a different texture and form than the sand. We would see stories in the snow where a hawk or owl might have taken prey from the snow tunnels that crisscrossed the grassy tufts. We would try to guess what had happened and imagine the drama of the night we saw imprinted in the snow. Our day would close with a visit to one of the local places that served clam cakes and chowder. It was always wonderful to come into the warmth after our adventure in the cold.

For many years, I was a member of the Silo Concert Dancers, a repertory dance company under the direction of Ernestine Stodelle. I attended her adult women's class soon after the birth of Steve to see if dance exercise would alleviate my sciatica. After several years of practice, I was invited to join the company, and had the opportunity to perform several pieces Ernestine recreated from the original Doris Humphrey repertory. I joined the performing company when she was recreating Doris Humphrey's *Water Study*. This is a dance performed by twelve dancers moving together as a body to represent the movement of the sea from calm ripples through a clashing storm and spray, and back to the rolling waves as they come into the shore. She sent us all off to the beach to study the waves and absorb the images into feelings.

Water Study begins in a kneeling curl, and the movement transfers in succession from one dancer to the next across the floor. As I knelt in my position second from the end, breathing in cadence with my fellow dancers, I would send my mind back to the shore, and recapture the rhythm of the waves. As we each took the movement from the dancer beside us, the wave propagated through us from one side of the room to the other and back. Each iteration changed slightly, just as each wave is a variant of the one before, but not a perfect replica. Performed in pale beige leotards and tights, lit in pink and blue light, the illusion on the stage of the surge and crest of the sea came to life. I had the joy of sharing a performance with my daughter Jenn when she was also in the company. It was a connection where she bonded with Ernestine as fellow artists, creators of the mood and theatrical drama through light and space. And we shared the recreation of the movement of the waves on the shore recaptured from our times on the beach.

Dancing for Ernestine was one of the greatest gifts of my life. I hold within me to this day the impetus to move in harmony with Nature, to express my deepest emotions, joy, sorrow, fear, anger, and longing in resonance with movement in natural spaces. The weekend before my mastectomy, I danced nude in my garden to commune with the golden October trees and to lift my arms and bald head to the sky, joyful to be alive, and defiant of the fear facing the unknown fate of my body a week hence. I drew strength from the endless surge of the ocean and dared to dance down my fear of cancer.

1.7 SATURDAY SOUP

The struggle between the need to become a good wife and mother presiding over a smoothly run household as my traditional roots prescribed and the urge to follow my own professional development path kept me in a constant state of unrest. In the middle of my post-doctoral work first with Winnifred Doan in Dr. Poulson's lab at Yale, then with Leah Lowenstein at the Boston University School of Medicine Department of Biochemistry, I saw how women pursued professional excellence in a male-dominated world. Both had a considerable support system of household help including, in Leah's case, a live-in housekeeper who took care of her boys.

Balancing work, household, dancing and children consumed a large space of my time in Connecticut. In this period of my life, Saturdays were special, but also fraught with overlapping obligations. I was the driver for children's dance class, karate class, grocery shopping, dry cleaning, and banking. In between, somehow the house had to be cleaned from time to time, laundry had to happen, and pets attended to. Family meals became a major production. As the children grew older, they took on preparing a meal once or twice a week, and on Sunday we had obligatory family dinner together after church even if friends came too.

But Saturdays had a special routine. I would get up early and make bread, usually two or three loaves of whole wheat with honey and sometimes raisins and cinnamon. I put a soup together in the large stock pot and left it on low when we were out running around. By the time we came home, the house would be filled with the scent of soup to be served with fresh bread and salad. There was usually enough for the various hangers-on who gathered at our house. Soup has the quality of offering comfort and the flexibility to reflect the bounty of the seasons. Whether it was a thick stew that started with beef bones, or chicken soup with lots of vegetables and rice, or a thick butternut bisque, there was always the shared process of paring the vegetables and browning the base for the stock.

In the growing season, Saturdays included a trip to Sunny Acres where we bought fresh fruits and vegetables grown right in our own town. Barn cats would be running around there to amuse the children, and we brought a kitten home from one such excursion. It was a small grey-and-white creature we named "Tidbit," but he soon grew to be "Tubbo" from eating too much and supplementing his diet with mice and other things caught in the garden. We were frequent customers of Sunny Acres and Florence, the farm owner, would give Steve and Jenny samples—a few Bing cherries, a peach, an apple. This was also a community gathering place where we would meet neighbors and friends running similar errands. Sometimes impromptu exchanges of children happened to allow them more time to play together.

When Jenn and Steve were small, we had play group with two other mothers, Maryanna and Paulette, who also had a two-year-old and a baby each. The children would "help" make little *panuches* when I made bread. They learned to pat and roll the dough, sometimes using the floor or the chairs for that purpose. Those little lumps would come to me all black, sometimes with our dog Bootsie's hair embedded in them. I deftly replaced them all with smooth clean ones with added raisins as they were rising. When the rolls were ready, into the oven they went with the initials of each child pricked into the top, just as the ones they had given me were marked. The children would come by from their playing to peer into the oven window and see them rise and cook until they were all brown and fragrant. Every time I make *panuche*, I remember those little hands and the joy they had in sharing something they made all by themselves.

As they grew up, they each learned to make bread on their own. The *cavale* we prepared at Easter and Christmas was a particular challenge. I remember my own first adventures when the dough ate my hands and I had to be rescued by Mother. Jenn finally mastered the art of keeping her hands coated with flour and working the dough to a smooth elastic finish. There is something soothing about the process of kneading bread. I always like to have Italian songs accompany me as I work. Nona said the music came into the bread to make it rise better. In her case, she sang in her little Italian voice. Neither she nor I were good at carrying a tune, but the music established a bond between us solidified by laughter and stories.

I realized through this skill sharing that it is really the women who carry the culture. I was fortunate that I had time to spend with Nona learning the ways of making pasta from scratch for ravioli, bread, and really excellent spaghetti sauce with meatballs. My own mother gave me the duty and also the freedom to explore cooking as a service and an art form. She went back to work full time when I was in seventh grade, and I was left with detailed directions for starting dinner for the family. Oh, we had some famous disasters during those days! No matter what, my Father said grace over the meal and enjoyed every bite. Even the scalloped potatoes that were gray from having a tablespoon of pepper on *each* layer, rather than one tablespoon for the whole casserole. For me, making a meal for the family was a grounding experience, one that gave me the confidence and satisfaction of being a meaningful contributor to the family. I tried hard to transfer that joy of sharing meals to my children.

My house was always open to our teens, even after my divorce. They gathered knowing they would be safe and well fed at my house. In the time when I was re-plastering the walls and tiling the bathrooms, I would have kids stop by and help, learning how to spread vats of joint compound with a three-inch trowel over the walls. There was always soup and bread to share, and workers were rewarded by having played a part in the success of the task. When I was in the Silo Concert Dancers performing troupe, class and rehearsals occurred every Monday, Wednesday, and Friday from 5:30 to 7:00 p.m., sometimes longer. So, the children had to take responsibility for cooking on Monday and Wednesday, and on Fridays we had pizza. I reflect sometimes that I was selfish to pursue dancing when I had a family at home, and I usually went directly to class from work, which meant I had not been home all day. When I was a teen, I was responsible to cook for the family because my mother was working, and we had dinner promptly at 6:00 p.m., no matter what.

Both my children cook well and enjoy the experience as a shared process, so I guess the unconventional obligations I put on them had a beneficial effect. They learned to be responsible for their own actions and knew that I trusted them to make good choices. I expected them to take charge of their own lives and hoped to give them confidence in their own abilities rather than depend on their parents for every decision. I still enjoy feeding people and take joy in sharing a meal embellished with stories, experiences, and sometimes tears. It's a gift shared across time.

1.8 THE LEGACY OF A TIMELESS ROMANCE

On February 18, 2018, both my daughter Jennifer and I received diagnoses of breast cancer. On this bleak February day, an unnaturally warm rain pelted the frozen ground. News of yet another school slaughter added an exclamation point of tragedy to the invasion of disease into the most personal spaces of family. An endless but un-heralded litany of misery assaulted me as Nature screamed for protection from unbridled greed on every front. The pit of despair loomed depthless and dark, luring the weak, the exhausted, the desperate into the abyss of drugged stupor or worse yet, the oblivion of deliberate ignorance. I stepped carefully along its edge to seek a way forward through this challenge.

In the midst of my sadness and near despair at facing breast cancer along with my dear daughter, my sister sent me this poem of hope and love she found in our father's war journals. He was a young soldier at camp, knowing he was facing a high possibility of going to war and not returning. He wrote to his sweetheart a vision of what could be.

Forever Valentine
Poem by Michael A. De Marco for
Marcella Strutzel
Written in 1943 while in training at
Camp David, North Carolina

There are some days in February, few
Rare, precious days when spring comes back and brings
Sunshine, breaths of warm air, and thawing winds.

If one lost track of time and left out March,
It would be Spring and easy to believe
That winter's gone and Spring is here to stay.
It was Spring! We two forgot it was not May
And went to see the tree, our tree, the one
By us imagined, planted in the grove
Where it was growing with the other maples.
And as we walked, the spongey earth and sod
Made squeaky, sucking noises when we stepped.
We did not care about the wet, it was
So good to see that all the snow had gone,
That all the rocks were drying brown and grey,
And that in low spots, clumps of fresh, new grass
Thrust straight, green blades through last year's rotted mat.

We walked on the path's upper side, where sod
Grew thick and made a lapping rim, until
It wound itself away as if it were
A spool of yarn that had unrolled and left
Us standing, when unraveled, at the root
Of our adopted maple. There it was
Growing up, big, spreading graceful branches
So its leaves would brush the other trees,
Cast shadows to protect from the hot sun
Dew drops, ferns, lilies, moss and violets.
I drew up close and rubbed the bark to see
If I could feel its pulse with my four fingers.
They touched a smooth spot just beneath the crotch.
There were no scars or rings; it was clean bark.
This was the spot for it I knew.

I took
My knife out from my pocket, opened it
And whet the blade on sandstone rock I found.
I tested the new edge for sharpness then.
It felt keen when I stroked it with my thumb.
The setting sun was perfect light, as if

I had commanded it to send its rays
At that one angle for my work, I dug
The knife blade in; and it cut quick and deep,
Clean through the bark to strike the white heart-wood.
Then with a turn I sliced a curve and made
The bottom of the point. Another twist
And I had drawn the other side. Now for
The cutting out and stripping of the bark.

She stood aside and watched my bright knife blade,
More interested in the sunlight's flash,
Play, dance of fiery pinpoints on the steel.
At last I turned to her, said it was done;
But for a moment she stood silent, pointing
With forefinger at the sun and sky.
A cool wind blew the clouds until they rolled
Themselves into soft balls of eider down,
Lined up and formed in banks of threes and fours,
Then slowly rolled away on melting bars
Of sunlight, revolving puffs of golden mist.
And as the sun dropped down to rest behind
The shoulders of black hills, the wind whispered
Goodnight and stretched cloud banners there to mark
With streamers red where the sun slept till dawn.
We looked and saw the beauty of sunset;

Then with an easy motion she swung round
And gazed upon the heart that I had cut
Deep in the clear, grey bark of the young maple.
The outline on the inner heart-wood clean,
White, shaped a solid fullness in the bark.
A rounded heart, and through its middle came
The arrow straight and slim, perfect and true,
As if struck there, half-way exactly by
The golden bow of Cupid letting loose
The arrow of true love deep in our heart;
A last sun's ray turned it to sudden gold,

Just for an instant made it livid, burning,
Shimmering and alive with quivering light.
Then it paled white, and slowly darkened
Silver, like the rising moon of coming night.

Our heart will grow and strengthen with the tree,
And with the bark become more warm and round.
Perhaps it may stir, beat, pulse like our own
Some day when sap starts through in Spring . . . when sun,
South wind, blue sky bring with them miracles,

My father was First Lieutenant Michael A. DeMarco, assigned to the OSS Special Reconnaissance Regiment, Company B under General Donovan in World War II. Their mission, coded PAT in May 1944, called for parachuting 15 men into the Tarn in France with orders to "harass and destroy the enemy, cut German communications and supply routes, and strengthen the resistance movement." Meredith Wheeler researched the history of the PAT Mission through the support of a Fulbright Scholarship. As I read this history again, the words that send shivers through me to this day are, "*Within two weeks, the south Tarn was liberated. Some 4,500 Wehrmacht soldiers surrendered to twelve OSS men and a few hundred Resistance fighters—most of them poorly-armed, under-trained maquis.*"[4] My parents named me Patricia in honor of this mission.

I was born in the "baby boom" following the end of World War II—a hopeful declaration by my parents that the world could still hold love, grace, and beauty. Father never spoke of his time in the War, even when we children would ask about it. The book was closed, and the scars of his experience haunted him until the end of his life. Yet, he served in the United States Information Service (now part of the CIA), and in the U.S. Foreign Service with dedication and commitment to build an America that would fulfill the promise offered to all immigrants and citizens. As a first-generation Italian/American, I have carried that same commitment to public service for most of my own life. The ideal that government serves the collective public interest and protects the weak and vulnerable from the tyranny of self-interested power has always driven my own personal and professional decisions.

The generation that fought together in World War II shared a bond of a common determination to stare down evil while standing on the moral high ground of integrity. Service above self, to the ultimate sacrifice of life itself, bound together the citizen-soldiers of that time, and established the conditions that built the greatness of America as a world leader and model for progress. But, in the aftermath of that war, the spirit of cooperation in governance, in institutions, in aspirations began a gradual decade-by-decade erosion. Today, Father would not recognize the America he fought for, and the prevailing government policies would shock his sensibilities to the core.

The sense of making life better for our children, the sense of making life better for *everyone*, has evaporated into a governance framework driven by corporate interests. Business and government have fundamentally different objectives. The special interests of multi-national corporations now drive public policy to the detriment of the health and welfare of the people as a whole. Tax and financial policies have disproportionately skewed the distribution of wealth to 5% of the populace, leaving more and more people in the clutches of poverty, even if they are working full time or have multiple jobs.

The system is designed by and for people who make money from the returns on their invested money. Corporate profits are at an all-time high, while wages stagnate or fall. Working hard does not guarantee success, or even a viable life. The poverty in America is a deliberate political decision. Likewise, the assumption that clean air and fresh water in American are guaranteed is fading. As pollution runs rampant while regulatory controls are rolled back, rescinded, or unenforced, millions of Americans suffer while living in polluted air and drinking unsafe water. Contamination from industrial operations disproportionately affects communities of color and people who cannot afford to move away. The environmental injustice rubs salt in the wound of having to live in unhealthy places, with no recourse or hope of escape.

Unlike the specific, horrific crimes of Nazi Germany, the slow violence of corporate greed raises few objections. The country increasingly is divided over ideology, politics, race, and religion. There is no sense of urgency to move in a collective effort to preserve a fair, equitable, and healthy future for our children. Everything rests on short-term benefits. There is no sense of action to create better options for our children. Any policies that purport to curtail the "rights" of individuals or corporations to profit, regardless of the consequences, are viewed with derision and denounced as burdens on business. What of the burdens on the next generation? What of the obligation to protect the innocent and help the indigent? Where is our higher calling to improve the community in which we live?

As the world continues to deteriorate, it's now necessary for everyday people to take up the mantle of reform with the moral conviction to make things better. It is imperative that people learn from the brave men and women who laid down their lives for justice, freedom, and respect for human dignity. The rampant racism underlying many of the current policies in America must be called out and condemned as the precursors to tyranny. Democracy is not automatically viable. It requires active participation by an informed and caring citizenry. We must be willing to stand and fight for the dignity and respect of all people, for the right for life to exist as intact living systems that serve our earth, for the fair and equitable distribution of the nation's wealth, and for the rights of people to express their opinions openly, and for respect when they do.

I thank Father and his many comrades in arms who came home from serving our country and left a legacy of hope for the future. We shared a spiritual connection through the roses that graced the garden in the front of our home. Father always sent me roses for my birthday and celebrations, and to mark special events. He taught me to prune and graft, and he mulched his roses with cocoa mulch he somehow obtained from Hershey, Pennsylvania. My favorite was the American Beauty rose, deep dark red and velvet soft, with a fragrance that scented the whole room when put in a vase indoors. In the early morning, dew would sit on the petals like sparkling diamonds. When the rose opened to full bloom, the yellow pollen circle welcomed the honeybees and later the rose hips formed to offer the birds a treat. Even in times of terror and horrific disasters, roses bloom and perfume our lives with a fragrance of hope and joy. Father saw me as the expression of hope for the future. It is a legacy I carry with humility and fierce dedication to the ideals for which he named me.

As I contemplated the bleak prospect of my future as a cancer patient, my mind wandered to the conditions my parents faced in a nation on the brink of war that had already ravaged Europe, with unknown perils looming ahead. My Italian-born Father, with his poet's heart yet unbattered by the horrors he would experience as a special service paratrooper in Donovan's Devils unit behind the lines, composed his wonderful anthem of love and hope for his fiancé. Their dedication to the fight for justice and goodness never ceased until their ways parted when Mother died in March 2000, followed by Father in April 2001. The love bond encircled me, my brother and sister, the extended family as it grew, the students they each taught, and the many colleagues and friends they acquired all around the world. They did not fight for people to bear arms, but for people to know and speak out, to stand against tyranny and injustice, whether on the battle line in time of war or on the picket line in times of strife over worker's rights. Theirs was a romance tested by time and adversity, a lasting legacy of love.

I did not know of my father's poetry until after his passing. The war had sent it into hiding. Through it, he left us a vision of hope for a better life that gives strength to carry on in tough times. His letters always began with descriptions of the gardens and the state of the fruit and the roses which served as grounding, beautiful thoughts in the midst of ugliness. Though we often clashed as I forged my own independence, we always connected through the roses we cultivated together in the garden. In the darkest of times, it is good to remember that beauty surrounds us in the simple expressions of nature. Opening our eyes to let in the light can bring hope to our hearts.

PART II

THE AWE AND POWER OF NATURE

Photo credit: Alaska Fish & Game Commission

". . . man does not live apart from the world; he lives in the midst of a complex, dynamic interplay of physical, chemical, and biological forces, and between him and this environment are continuing, never ending interactions."
- Rachel Carson, *Lost Woods*, p. 228

Part Two

I divorced my first husband when both children had moved off to college to begin their own lives. I found life as a divorcee lonely and isolating, and ended up marrying an Alaska fishing guide, Joe Barkoski, and set out with his dog Homer to live in Anchorage, Alaska where I stayed for ten years. During that time, my career rose to its pinnacle as I was president of the Anchorage Economic Development Corporation then a commissioner of the Regulatory Commission of Alaska and finally an associate dean of the University of Alaska College of Business and Public Policy. I suffered from uterine and fallopian cancer and was treated while serving as a commissioner of the Regulatory Commission of Alaska. As a civic leader, I was surrounded by many friends in this difficult period. Both of my physicians were personal friends. I had a strong circle of women—the "Sans Uteri"—who gave me strength and kept the fun in life alive. The essays in Part Two are from this period.

2.1 THE AWE AND POWER OF NATURE

In 1992, my life took a sharp change in direction. My marriage and life as a doctor's wife stretched before me void of promise beyond the inevitable progression from youth to age. I felt a constantly simmering anger at my professional standing, always the brains behind the male figures in the scene, often the wordsmith of their proclamations, but denied a place as standard bearer. One day, I noticed an asymmetry in my neck, just above my collarbone—there was a swelling on the right side. It turned out to be a thyroid tumor, benign and not invasive, and surgically removed. I coordinated the Connecticut Energy Advisory Board conference, attending to the logistics of speaker's briefings, breakout room assignments, and registration and program coordination, all with a turtleneck sweater covering my bandage. The diagnosis came with the cause: "chronic acute stress."

I despaired to the point of planning to take my own life. I sat writing tearful notes to my parents, my children, my husband—the list kept getting longer. Outside my study window, a chickadee sat scolding me by hopping up and down because the feeder was empty again. I looked into the heavy rhododendron bush shading the window and watched that insolent little puff of feathers challenge me while I wallowed in self-pity. I tore up my farewell notes and began to lay out a plan for my ideal life.

My children had gone off to college, I was divorced from my husband, and experienced the shock of being single in a coupled world. My work was progressing with the experience of launching TECHCONN with two outstanding colleagues. But instead of having the three of us who had built the organization and landed a $5,000,000 ARPA grant in the leadership positions, politically connected men displaced from the shutdown of Electric Boat in Groton were hired. My colleagues in this project, Mike and Cliff, went off on their own business ventures, and I went to Alaska with a displaced General Motors machinist aspiring to be a fishing guide. He courted me with roses, rides on his motorcycle, and an adorable Labrador retriever puppy named Homer, after Homer, Alaska. I rented out my house, stored my furniture in Pittsburgh with my parents and set off for the Last Frontier.

From 1995 to 2005, I lived in Alaska where the lieutenant governor, president of the senate, and speaker of the house were all strong women, and many of the Alaska Native communities are matriarchal in structure with a tradition of women elders as the principal force. Alaska broke all barriers for me. The glass ceiling cracked for my career as I took my place leading the Anchorage Economic Development Corporation and left the legacy of the Economic Engines of Anchorage plan. I was a Commissioner of the Regulatory Commission of Alaska during the reorganization of the agency, and finally joined the University of Alaska-Anchorage College of Business and Public Policy as an associate dean. My career had reached a pinnacle of achievement I never would have imagined or achieved in Connecticut. I had the thrill of serving on the board of the Anchorage Rotary Club and the Anchorage Symphony Orchestra. My dinner parties became legendary meeting places for visiting scholars, artists, and musicians. I learned to fish but declined to hunt.

Soon after I was appointed a commissioner of the Regulatory Commission of Alaska, I was diagnosed with both uterine and fallopian carcinomas. In the case of the fallopian tumor, its discovery as part of my regular annual gynecological exam came weeks before the tumor would have penetrated the wall of my ligated fallopian tubes. Treatment at that stage is usually futile. Following my full hysterectomy surgery, I spent six weeks staying as still as possible on the edge of Oney Pond in Anchorage where I was living at the time. Creatures came to me there and shared their secrets. Seeing the amazing manifestations of the triumph of Nature in Alaska sustained me to prevail over the illness that sapped my strength.

In the midst of the heady swirl of world travel to represent Alaska and trips to many remote parts of the state of Alaska itself, I gained confidence as a speaker, and learned many insights from the people I met and worked with. The community of close friends embraced me through the horrible experience of having uterine and fallopian cancer while in a public position. Like Rachel Carson, I needed to conceal my illness as much as possible to avoid undermining the decisions made during that time. The Trans-Alaska Pipeline Tariff was being challenged, and any opportunity to discredit our review would be exploited by the oil companies who had a major interest in the case. The magic of Alaska sustained my spirit even as my body failed.

In this decade of close experience with the frailty of life, I was privileged to have some of the most spectacular encounters with the wonders of Nature. The resilience of the creatures and plants of this great state gave me hope. In writing my journals, I found healing and celebration in the natural world made precious as I recognized how fragile my own life had become. In the long hours of having chemotherapy and in the tortured hours of night when my body ached and provided a challenge to rest, my mind took refuge in the memories of the wonders thriving all around me. Let me share some of those scenes that lifted my spirits and allowed me to find refuge from the pain.

2.2 SCENES FROM ONEY CIRCLE

My house on Oney Circle was situated just below the tree line, a short distance from the Chugach National Forest. At the time I lived there, the nearest house was a half mile away. The house was a log structure with the entry on the ground floor and the living space upstairs with a view of the Anchorage Bowl and Denali beyond. There was a small pond fed by the mountain stream and a dock that floated only in the fullest times of the spring snow melt. On many of the evenings and afternoons when I was recovering from my surgery and too uncomfortable to sleep, I would count the triangle shapes created by the wood panels of the ceiling and listen to the sounds of the night.

Aurora Borealis

On Christmas Eve close to midnight, I was walking in the quiet calm of new snow under a sky clear between storms. All the lamp posts on the driveway were glazed with icy frost, their light diffused into a soft glow. The spotlight over the driveway was out for some reason and the night was dark in a cloudless sky with no moon. Many of the neighbors were Outside (out of Alaska) for the holidays, their houses invisible in the darkness. All the constellations of heaven sparkled in the black sky, seemingly only an arm's length away—Orion the Hunter, Cassiopea, and the Great Dipper, right above the house. Here on the edge of a mountain close to the tree line, the heavens felt so close. I remembered my Fathers tales of the sky from my earliest childhood days. I strolled in wonder down the familiar driveway, marveling at each alder bush covered with bright ice from the afternoon rain that preceded the light snow. No purchased decorations could rival the iced alder seed cones.

I found two young moose bedded down in the snowbank behind the house. They curled up beside each other and their heat melted a little cave into the snow to protect them from the wind. They often slept against the house under the vent from the furnace. In the early morning, they would munch my bushes for breakfast. But on Christmas morning, I could not begrudge them their meal.

It was hard to be away from family at such a time. Christmas memories brought a tightness to my throat as I imagined the whole body of the 58 relatives and friends gathered in my parents' house celebrating the *Vigilia* of Christmas. For so many years I was there to clean the squid and make the *cavale* and *pizzelle* with Nona to serve the family gathered on Christmas Eve. I knew there would be celebrations in Alaska, but I was not yet knit into the community. I could not help but feel a bit sad and homesick as I wondered about the wisdom of moving so far from my own roots. The sharp cold of the Alaska night pierced me to my bones and the world felt hard and harsh.

As I reached the cul-de-sac at the end of the driveway to let my black Labrador Homer run around, the sky was suddenly illuminated with a display of the aurora borealis. The ribbons of green, pink, and yellow cascaded toward the horizon like lightning in slow motion. I stood there well wrapped in my fur coat and hat but shivering in wonder at this majestic display. Each tree projected an iced white silhouette against the moonless black as the aurora borealis continued. Watching Nature's light show without the drama of thunder and lightning, precious in its rare appearance so far south, I felt a welcome to this place, Alaska, that shook me to my soul. Here was a place at the top of the world where my soul could commune with the Spirit of the Earth up close.

So much in Alaska is on an enormous scale compared to other places. The sheer vastness of the state beckons exploration, yet there are fewer paved roads than in Rhode Island. I vowed to explore as much of it as I could physically manage. My resolve to feel a part of this place of mystery brought comfort to my heart.

Duck Talk

Nestled in a copse of alder rimmed with tall spruce that towered above the log house was Oney Pond. It was about an acre of freshwater pond fed by intermittent streams that came down the mountain, gushing after heavy rain, reduced to a trickle in the late summer. The pond was a perfect duck habitat with willow trees overhanging one edge and a marshy wetland at the other. It was deep enough in the center to shelter fish and the edges are graced with water lilies, arrowleaf, and grasses.

Spring came late to Oney Pond this year, and in the afternoon just before sunset, a pair of mallards landed on the thinning ice and skidded to a stop. Instead of the open water they were expecting, the pond stretched frozen and hard almost to the edge of the inlet. They took up a grumpy residence under the picnic table, and I threw sunflower seeds out to them. I noticed that the duck had a lame left leg, perhaps an injury inflicted in escaping from a predator in the wintering grounds. I worried about how she would fare as a mother with her lame leg. The mallard pair was joined over the next weeks by two more mallard pairs and a group of bachelor drakes, three pairs of pintails, two sets of widgeons, and a barrows golden eye.

There was a raft with a hutch on it anchored in the middle of the pond. Here the various ducks would sit and sun safe from land predators but exposed to the eagles unless they went into fast dives. When the eagle was perched on top of the spruce tree, the ducks all gathered in a squawking huddle and flapped their wings in alarm. One morning in May as I sat on the dock having my coffee, I noticed a widgeon with a stream of ducklings in her wake. I love the widgeons with their taupe feathers and beautiful dark eyes. The brown and yellow mallard ducklings are distinguished by the stripe over the eye and are nearly indistinguishable from pintail ducklings that are more grey in down than yellow. All the mama ducks kept their little broods together at first, so it was relatively easy to count them. At one point, there were 75 ducklings scurrying around the pond.

As the ducklings gained confidence, they separated into the currents and swirls of the inlet, probing among the grasses and water lilies. The lame mallard was a very protective mother. She did not tolerate her ducklings wandering off alone but chased them down and kept them close. She also chased away other ducklings and mothers if they came too close to an area where her duckies were feeding. She liked the edge of the marshy grasses where there were lots of insects, leeches, and little fish. There would be frequent squawking matches among the mother ducks as they vied for prime feeding spots.

It was my custom at the end of the workday to get into my canoe, usually with my dog Homer in the bow for ballast, and paddle around the pond. I would scatter sunflower seeds so the ducks would come and follow in my wake. I soon learned their little quirks and personalities. They had a pecking order where the mallards chased the widgeons, and the pintails challenged the mallards for space on the edge of the duck raft. I had to tie Homer into the canoe so he would not jump in the pond and chase the ducks. He would wag his tail briskly and bark at them, but they soon learned he was confined to the canoe and paid him no attention.

I was sitting at the picnic table sorting through my mail one afternoon when a huge moose sauntered to the edge of the pond. She munched a few water lilies, flowers and all, then took a leisurely swim across the pond. The ducks all came to trail in her wake snapping up the morsels driven to the surface by her passing. She emerged at the other side, shook off the water and made her way into the alder copse and on down the hill. The ducks were very busy that afternoon enjoying the bounty she had stirred up in her passing.

One morning, I heard the eagle screaming from the spruce tree. I rushed outside to see it swoop down, circle the pond to drive the ducks all into the middle, and dive! She rose into the air with a duckling in each claw, headed up the mountain to feed her own nestlings. I watched in sadness as the eagles came back again and again to take ducklings to feed their own brood. It would be weeks before there would be salmon in Ship Creek to lure them away. The ducks mounted guard squawking and flapping in close groups, and the little ducklings would dive deep. They were safe from the pounce if their dive was timed well.

The widgeons practiced a different strategy. When the eagles were about, the mother widgeon collected her brood under an overhanging willow and made them nestle under her wings. Then she sat very still and tried to look like part of a branch in the water. This proved to be an excellent strategy. By the end of the raids, there were only 17 ducklings left in the pond to grow up, practice flying, and leave for the winter.

Usually the pond was a peaceful place. I looked forward to my afternoon paddles around this small domain. I came home one day to find the lame mother mallard walking back and forth at the edge of the dock in a distressed manner. As I cautiously approached, I could see that a little duckling had fallen off the dock right into the canoe. It was too small to hop out or to get onto the bench and make its way to the edge. It was scuttling back and forth in the few inches of water in the bottom of the canoe peeping for its mother. I sat on the edge of the dock and gently scooped the little thing into my hands and released it into the water. It joined the six ducklings waiting there herded by their mother into the shallows beyond the dock. I turned my canoe over and rested it upside down on the dock to keep the ducklings safe from further entrapment.

I loved the Barrows Golden Eye ducks. They were quite striking with white dots along their black wings and slightly crested heads. When they took to the air, their feathers made a unique whirring sound. I was excited to have them nesting at my pond. They usually nest 100 feet or so away from the edge of the water. Their nest was on the uphill side of the pond deep in the alder brush. It was late at night, one or two in the morning and quite dark, when I heard a commotion outside. I went to look with the large flashlight to see what was going on. I heard nothing more, but saw the great horned owl rise against the sky. In the morning, the Barrows Golden Eye was alone. She made a slow circuit of the pond and flew away. I ventured into the alders to see what I could discover and there was the sad tale. The nest had been raided, all the ducklings gone, with only a few down feathers strewn about.

There is nothing more dramatic than the cycles of life in Alaska. The predators and prey coexist in close quarters with humans. The small dramas of daily life go unnoticed in the cities, but here where the houses are sparse and overshadowed by the trees and brush land, the daily drama of the cycles of nature play out close at hand. My empathetic emotion contributes nothing to the cycle of life. It is amazing to see that nature provides balance, even here in this small, isolated pond. All 75 ducklings could not grow up and thrive year after year. More than double the number of parents survived to fly away this year. I wonder how many and which of them will live to return next year after a winter of adventures in the South.

The Still Point

In the weeks immediately following my surgery for fallopian and uterine cancer and before the weeks of chemotherapy that followed, I would sit on the edge of Oney Pond, wrapped in an afghan holding a hot cup of tea and allowing the world to settle around me as I tried not to move at all, even to breathe. Often engulfed in the fog of pain killers, I would sit in a half-doze and try to absorb the stillness around me.

In my state of blending into the background, the little creatures ignored my presence and went about their business. I would put a heap of sunflower seeds out and the birds would come to feed. They came so close I could hear the wing beats in the air and the chickadees would fuss and bicker over their chosen seeds. The little ermine was turning white in anticipation of winter, and he snuck around the edges of the dock. The ducks were fledged into winter feathers, and some were practicing their first flight lessons across the pond.

It was a partly sunny afternoon, warm for September. There was no wind and the stillness was so complete, I could almost feel the mountain breathing. I noticed movement at the edge of the pond where the alder brush is thickest so I focused my attention there. Out of the shadows crept a mother lynx coming down to the water. She was wary and alert to any movement as she slowly lowered to the edge of the pond. She kept her eyes lifted as she lapped the water. I could hear the sound of her drinking and saw the ripples spread across the still water.

Beautiful in her wildness, she was facing a daunting winter with her cubs as the snowshoe hare cycle was beginning its decline. I did not see her cubs that afternoon, but I saw the brush rustle as she returned up the slope to where they were hiding. Of all the creatures I encountered by accident in Alaska, seeing this mother lynx in the middle of a residential area declared the wildness had not yet been conquered here.

The ripples of that moment stayed with me through many afternoons after work when I sat in the chemotherapy room, sometimes long into the evening because the infusion took hours. I would think about the trials that mother lynx would face as the winter advanced. Months later after a snowfall, I was walking the dog early in the morning, barely at sunrise and I saw the tracks in the snow of a lynx and one cub. I wondered whether this was the mother lynx I had seen before, and if she had lost one cub already. Or was this perhaps another pair that had come to hunt in the alder brush for the winter hare?

Ice Fog Follies

Single digit temperatures for days on end drove me into fur and hibernation. The air was clear and crisp in the mountain house. The short-tailed red squirrel scurried to and from his den with cheek pouches stuffed with sunflower seeds stolen from the bird feeder. He scraped away at the door rug to take fuzz to line his nest against the biting chill.

The sun rising over the horizon turned the white and gray world into an infinite palette of pastels from pink through all the lavenders and mauve to blue at the top. Tree outlines seemed etched into the skyline. The Anchorage Bowl was invisible under the cloud of fog with the Chugach Mountains crowned in bright pink and magenta as the sun touched their snow-covered peaks and threw Denali's stately crest into a bath of frosty light 300 miles away.

My day's pursuits took me below into the fog, into the city, into the stoplight follies. Here the ice fog coated everything with a half inch of crystals . . . the roads, the vehicles, the houses, each branch and twiglet, the power lines, the sidewalks. As cars idled at stoplights, the condensation from their exhaust froze and glazed the intersections with diabolical treachery. The approach to each intersection felt like a game of roulette. Will the oncoming car in the cross-street brake gently and soon enough to stop, or will the vehicle come plowing toward me as I cross the intersection, wheels locked in a sickening skid, out of control, into my path? As I sat at each stoplight, I watched with apprehension as a vehicle pulled up behind me. Will the driver leave enough room? Will there be space to dodge out of the way if there is a skid into my rear bumper? As I drove along, I wondered whether each little patch of clear road might really be glare ice in disguise as a safe place.

At noontime, snug in the safety of my fur and boots, I could enjoy the more spectacular attributes of the ice fog without the terror of driving. As I walked along the sidewalk, with snow banked to my shoulders on each side, I could see closely the effects of the ice fog. Each tree became a fantasy caricature of itself. The layers of ice crystals glistened like a million jewels in the sunlight. The little birds that flew through the branches shed a spray of glittering ice as they darted in and out. Light fractured into rainbows filled all the shadows with new possibilities.

In the evening as I drove into the lane at home, the coach lanterns along the driveway caused reflections in the ice on the snow. Icy jewels were strewn with the largesse of a great king over the entire landscape. The footprints of the browsing moose showed as deep blue shadowed indentations in my driveway. The moose were bedded down in the alder for the evening, one in the front yard, and one in the side yard near the library window. I wondered if they too would be covered in ice crystals by morning.

2.3 MOOSE MADNESS

Living in Anchorage on the mountain above the city brought daily life into regular contact with the wild residents of the neighborhood. Throughout the year, there were moose all about. These are not animals to be trifled with or taken lightly. They are not placid cud chewers like dairy cows. They owned the place and did not tolerate interference from humans. They have no sense of humor at all.

One of my great pleasures was to hang my laundry out to dry in the sun and mountain air. I had a long clothesline stretched across the back yard to which I had pinned two sets of sheets and pillowcases to dry. I was washing dishes and looking out the window when I noticed a large bull moose strut into the yard. He was clearly annoyed, snorting and pawing at the ground. I soon realized that he was taking offense at the linens flapping on the clothesline. He charged at one of them and backed up to see the effect. The linens kept flapping away in the wind. Now, he was really annoyed. He charged right into the sheets, slashing out with his hoofs and trampling the offending material to the ground. He finally stomped off with his antlers adorned with the torn shreds of a pillowcase. The poor sheets were reduced to rags.

Moose are very protective of their young. One spring morning, I looked out my window to see a mother moose drinking at the pond. She had left her twin calves in the alder brush, but as she was not close to them, they had ventured out and were gamboling around on the sloped edge of the pond. One of them tumbled over and almost rolled to the water. The mother finished drinking and herded the two off into the woods to hide again.

I never knew when I would encounter a moose. There was a bend in my driveway about halfway to the road that curved around a large boulder. The moose liked to sleep there piled together against the boulder in the curve of the road sheltered from the wind. There were four moose curled together in a heap and not ready to get up. It was barely light out at that hour. It had snowed a little overnight, and each moose was dusted with ice over their thick fur. The problem was that they had completely blocked the driveway so I could not pass to go to work. I tried flashing the lights. I tried beeping the horn long and loud. One lifted its head and glared at me but made no motion to get up. Finally, I backed up and revved the engine and blared on the horn at the same time as I moved forward. Now I had a new problem. All the moose stood up, faced me, and started stomping and snorting at their rude awakening. I stayed in the car and kept beeping and they finally ambled off to the side. I was a half hour late to work that day.

One afternoon, I had a meeting at the BP building, which had a large entry way with an interior courtyard that had a fountain and luxurious indoor plantings. As I was preparing to leave the building, the security guard stopped me and others at the top of the escalator. A moose had entered the courtyard through the main doorway and was munching away on the decorative greenery, showing no interest in making his departure. Someone had summoned a veterinarian to bring a tranquilizer gun to remove the beast, but he was at least a half hour away. In the meantime, we could only wait and watch the moose destroy hundreds of dollars of plants.

It was also not uncommon to see moose walking through the city streets. They were particularly fond of ash trees and willow plants that grew everywhere. The moose showed no fear and were rarely challenged. It was so uniquely Alaska to see a full-antlered bull moose striding along in the median strip on the Northern Lights highway. I always maintained a sense of respect for these creatures who refused to give up their space to any human enterprises or presence.

Moose encounters were not always harmless. There were many instances of collisions with cars in which the moose and drivers did not fare well. Because of their long legs, a moose would tend to come through the windshield in a collision, with disastrous results. It was also a particularly dangerous time to encounter moose when mothers were with their young ones. They became fiercely defensive and would charge at people with almost no provocation. I was walking back home along the road from Prospect Heights Park with Homer on his leash when I came upon a small group of moose in an empty field near the road. One adult was partly in the road munching on a birch tree, a mother and two little ones were facing the road, not eating or moving—a very bad sign.

Since they had seen me, I really could not go forward to pass them. There was no way to detour around them, and I was afraid to turn my back and reverse my direction. So, I backed up slowly and tried in vain to keep Homer from jumping up and down and barking at them. While I was still in a fearful retreat, a pickup truck came around the bend toward me, and the driver saw the predicament I was in. He suggested that he would drive between me and the moose and I could pass behind the truck and make my way down the street. This little diversion was executed brilliantly. The moose stepped away into the field to get farther from the truck and I went past with Homer and ran down to the bottom of the street and on to my own driveway a block or two away.

Nothing was safe from the moose when they were hungry, which was all the time. I had delusions of planting a garden around the house as had been my custom in Connecticut. This was a disaster. What the rabbits did not munch, the moose devoured. I finally resorted to putting plants in hanging baskets high above the ground. I had started tuberous begonia bulbs in the house and finally hung them in baskets at the entryway and beside the swing overlooking the pond.

One day I was upstairs making dinner when the doorbell rang. I saw no vehicle in front of the house and had seen nobody coming up the driveway. I went down to the doorway and was puzzled to see the glass insert in the door blocked by a brown shaggy obstruction. Just then, the doorbell rang again. I was seeing the back end of a moose scratching his behind on my door jamb as he cheerfully ate the begonias hanging six feet in the air. I gave up on the gardens and only planted flowers and some vegetables in tubs on the upper deck. Only the birds could reach there, and I was able to grow basil, oregano, and lettuce as well as flowers all along the deck.

2.4 TIGHT LINES AND DREAMS

Since I married a fishing guide, I had many opportunities to engage in fishing in pristine areas. Fishing is a central part of the life of Alaskans, not only because of the economic base of the industry but also as a way of life. When I was at the Economic Development Corporation with offices overlooking Cook Inlet, there was a king salmon run that came through Ship Creek, right in the city. Two staff members brought in a huge cooler filled with ice and left it in the break room. They went off to lunch in their business suits, but with hip waders on, and returned with a king salmon apiece from an hour at Ship Creek.

When the fish were running at any particular hot spot, people would gather shoulder-to-shoulder flaying the water to a froth and battling over tangled lines and stolen fish. I dreaded such expeditions and stayed well out of the way. I much preferred the more solitary ventures where I could absorb the grandeur of the place and experience the sense of connection with wild creatures. Fishing in out of the way places for trout, arctic char, or silver salmon became a source of inspiration that filled me with gratitude for the sustenance provided to so many people and other inhabitants of the web of life.

Trophy Trout

I had been having some intense times at work for a few weeks, and Joe was without a fishing client for the first weekend in months. So, we went camping to Summit Lake, a place we passed all the time on the way down to the Kenai River. It had a small campground, only twenty sites, and was officially closed for the season. Only a few hardy bear hunters and a couple of kayakers were at the site. We arrived in late afternoon and had time for a quick paddle around the lake before dark. There were fish in only three places on the whole lake: where the Upper Lake flows in, where the lake itself drains into the stream that will eventually join the Kenai River, and at a freshet that runs down from the mountain. We each caught three small trout and I cooked them for dinner. There's nothing quite as good as freshly-caught trout almandine eaten by candlelight. After dinner, we enjoyed a stroll in the setting sun with Homer the Wonder Dog before turning in for the night. Summit Lake was still and silent with the fragrant air of autumn laced with campfire smoke. I heard the lilting call of a loon as I drifted off to sleep. The press and commotion of Anchorage were far away and out of mind.

We started out fishing early the next morning. It was a calm clear day, but the sun had not yet risen over the mountains. The boat slipped through the water making swirls in the mist as it passed. A pair of trumpeter swans glided in and out of the mist like visions in a mirage. The water was so clear and still, that the mountains golden with autumn trees and red from high bush cranberry reflected a mirror image along the far shore. We had turned back and forth once across the end of the lake with no results when we were greeted by an elderly gentleman in a kayak. The old man remarked that it was just another lovely day in paradise. I agreed with him but noted that it probably was not a good day for trout because of the sun and calm water. I was using my trusted eagle claw rod with a Shimano reel and four-pound test line with a little rapalla lure on the end simulating a stickleback. The boat was in a wide turn, and the rapalla trolled along the edge of a grass bed near the edge of the lake. Just then, as I finished my words, a fish struck my lure!

I pulled back hard, and a huge rainbow trout leaped high out of the water. The struggle was on in earnest. Joe knew it was a really big fish, and he coached me carefully so I would not put too much pressure on it because I had lightweight line. It leaped again and again and ran at the boat, and I had to reel fast. It tried to run under the boat, and Joe urged me to keep it well away from the side, so the line would not be passed over the edge of the boat or get in the motor. Finally, the fish came close enough over the bow that Joe was able to get it into a net.

The fish was spectacular in color, a female with greenish camouflage on her side fins. She had been lurking in the grass, watching for a little stickleback, when the rapalla floated by wobbling in the wake of the turning boat. It appeared to be easy prey, but deadly with multiple sharp hooks. I finally felt the absolute thrill of fighting a true wild trophy fish and landing it myself. I like trout so much better than salmon. They jump and fight and really have so much determination that I almost hate to keep one. I mostly let them go, except those we eat. But this one was a true trophy, and I knew I would cherish this day for a long time. I thought that when I hang the mount in my office, I would be closer to being a Sourdough Alaskan!

Roslyn River Afternoon

The Roslyn River on Kodiak Island offers a magical setting for fishing or just letting the sense of the place seep into my soul. The parking area is about a quarter mile from the water. The path runs through a cedar forest where all the ground, tree trunks, and branches are clothed in brilliant green moss. Ferns and wildflowers poke up in clumps on the carpeted way, but there is no brushy understory. The ground feels spongy when you step on it, and the green-on-green shade gives evidence of the Emerald Isle nickname for Kodiak. Even the air smells green with the heavy scent of thick, damp vegetation—like a jungle without the heat. When you reach the river itself, it winds its way through the overhanging branches with stretches of straight runs interspersed with deep pools. There the fish gather on their way upstream to spawn. You can walk all the way down to the mouth of the river where the surf rolls in over sandy beaches. The water is gin clear and not too cold.

I stood in neoprene waders about three feet deep in the incoming waves and watched the water for signs of fish. I knew they were there, because the harbor seal was busy, and he is an excellent fisherman. Every now and then a bright silver salmon would leap into the air and splash back. Suddenly, a fast moving and large group of fish was swimming around me. They were quick and bright, and bumped into me as they darted past. I was so startled, that I could hardly move, never mind actually cast to them. Then that bunch was gone, but another would come by soon. I watched for the ripples on the surface, or the swirls in the water where they would turn and make an eddy in their wake.

And then, in the curl of a wave as it gathered, I could see fish in the wave itself. I cast a lure right in front of the wave, and just as it was breaking against me, I felt the tug of a fish on the line. I set the hook with all my might and the fish took off back toward the open water. I let him run a bit, then slowly started walking backwards toward the shore. He shifted and ran right at me, and I had to reel fast to keep ahead of him. Then he zigged and zagged around me, trying to tangle me in my own line, but I turned and turned as he moved, and kept the line reeled in taut. Finally, I had the option to haul him the rest of the way onto shore or let him go on his way upriver to spawn. He was a beautiful fish, dark green on the back and silver on the bottom, with a blue eye. I pulled out my long nose pliers and carefully removed the hook from the corner of his mouth. He was gone in a flash.

I had managed to get my line all tangled while unhooking the fish, so I had a lot of it out into the water reeling in slowly to get rid of the tangles. As I was not paying much attention to my task, I thought my hook had seized on a seaweed clump on its way along the bottom and gave it an irritated jerk. The line took off with the drag singing in protest. I had accidentally hooked another fish. This one was not amused at all. It was a female, and she jumped out of the water again and again. She swam all around in circles and kept heading out into the waves. I had too much line out for a good fight, so I had to keep reeling hard to catch up. She had gone out and been hauled back four times before I began to gain ground. I did beach that one, because the hook was swallowed deep and not easily removed. I thanked the fish for her gift of life as I cleaned it and put it in ice. We would have fresh salmon for dinner.

I kept casting into the waves, but not really concentrating on the fish. I was too distracted with the fishing tactics of the seal. He would come up and peer around as though spotting through a periscope. Then suddenly, down he would dive, and there would be a lot of churning. I could see the ripple of fish as they sped away ahead of him, and sometimes, there would be a red splotch on the water, and the seal would float on his back eating for a while. Then the sea gulls would plague him for the leftovers, which were spreading out on the water. I was watching through my field glass, with my fishing rod stuck in my waders to keep the reel out of the water and the hook reeled in tight. I was unlikely to catch fish that way, but there was so much else to see there. The eagles also watched the action, because they knew there would be pieces of fish left over after people fillet their catch. They compete with the gulls over the entrails left in the water by all the fishermen.

I felt really a part of this scene, up to my middle in the rolling water with wild creatures swimming all around me, and some catching the others. There was no direct sunshine, only a foggy brightness with a golden cast to it. The spray from the ocean mixed with the condensation dripping from the clouds. It was a very strange, desolate feeling. There was only one other fisherman far down the shore catching pink salmon one per cast. He was either a tourist who didn't know the difference between pinks and silvers, or he was catching fish to feed his dogs for the winter.

This area was known for its strange lore. There was a tale of a Native woman who watched from the woods and set her eye on a likely fisherman. She would lure one into the river and take him for her own. It was told that long ago a woman had lost her husband to the sea, and she waited and watched for his return. The place was shrouded in soft mist and surrounded with ancient forests where the moss had grown thick, and the river cut deep into its banks. I had this pristine place to myself for an afternoon and marveled at the richness of life teeming there while I wondered at the mysteries of its past.

Halibut Queen

We were camping in Pasagchak Bay on Kodiak Island around the bend from the missile launch base. The water of the bay was as smooth as glass, not a wave in sight under a cloudy but calm sky. We set out in the boat with a bucket of smelly, brined herring for bait. We were heading along a cliff's edge just off a kelp bed. I was distracted watching a golden eagle coaching a fledgling to fly. The adult was holding a fish and hopping just out of reach until the fledge fluttered awkwardly to a branch lower on the cliff. Then the adult gave it the fish and went in search of another. I was a bit apprehensive about fishing since I had not done anything so strenuous since my abdominal surgery seven months before, and I was still quite apprehensive about the idea of fighting a fish of any size. But we were seeking salmon, and I knew that silver salmon were manageable, so I was trying to be brave. Our dog Homer was asleep in the sun resting in a pile of rope coiled in the bow of the boat.

Joe was calling out the water depth from the fish finder so we could adjust the level of the bait to lie just above the bottom. He called out, "30 feet, 35 feet, 42 feet, 45 feet, 51 feet, 40 feet. Lift your bait a bit." Well, I took a few turns on the line to lift my bait, and thought I was stuck on the bottom again, as usual. I was sure my double hook setup had stuck on the ledge, so I yanked on the line to get it loose, but it didn't budge. I grabbed the pole with both hands and pulled back with all my might. Nothing moved. I was complaining to Joe that I had lost another rig in the bottom, again. A moment later, my rod dipped sharply into the water. The reel's drag started screaming, and I thought I was going overboard. I had a fish on the line!

Joe patiently explained how I should move my left hand forward, support the rod under my arm, and reel with my right hand. He was very encouraging and supportive, but I was afraid to move my hands. There was no way I would be able to hold that rod with one hand and the reel with the other. I was protesting to Joe, and begging for help, but he was determined that I would land this fish myself. The sea was calm and the water clear. I held on with both hands as the boat turned, and saw a huge halibut undulating through the water, heading toward the open sea and right for Japan. This was a really big fish. We chased it out into the open water of the bay because there was no way to stop it until it became a bit tired from hauling the whole boat.

I reeled with all my might and kept the line tight to bring the fish up close enough to the boat that we had to watch it swim away again and again near the surface. Some thirty minutes later, we managed to haul it close to the boat. Joe yelled for me to pass him the harpoon. I had no idea what he was talking about but he finally explained that it was the aluminum rod attached to the rope with the buoy at the end. Well, Homer was still sitting on that, so I had to nudge him off while still holding on to the fishing rod. I managed to pass the piece to Joe, but I was totally unprepared for what happened next. He speared the aluminum harpoon into the halibut, and it took off again.

The rope to the harpoon was somewhat tangled,

but it played out smoothly followed by the bouncing buoy. I saw that the shaft of the harpoon was bent almost in half from the force of the thrust. The halibut went charging off towing the buoy and pulling out almost all of my line. We chased it with the boat to where the buoy submerged then bounced to the surface. Joe used the gaff to pick up the line from the buoy and with leather gloves hauled the halibut hand over hand up to the boat. I was still hanging on to the rod for dear life with no idea what was coming next.

Joe was standing ready with the "snake charmer" and took aim and hit the great fish behind the eye with a 0.4 slug. Blood streamed into the water leaving a spray red ribbons. I fell over into the bait bucket when the blast went off and stayed there trembling while Joe lashed the fish to the side of the boat. I had a distinct pain in my side and was totally exhausted from the fight and the trauma of the end. There was a definite list to the side as the fish was nearly as long as the boat.

In the meantime, three salmon sharks had joined the party, about a hundred feet away. They circled closer and closer as we slowly motored back toward shore. Joe fired into the water several times to keep the sharks away from the boat. We hauled the halibut out with the help of four men on shore and loaded it into a trailer usually used for hauling equipment and tents. We took it into town to check in at the Fish and Wildlife checkpoint. We lifted it with a block and tackle and found that it weighed 168 pounds and measured 67 1/2 inches long, bigger than I am. We spent the rest of the day carving up the fillets of fresh halibut and sealing them into the vacuum sealer.

A small crowd gathered when we were weighing in because none of the commercial fishing guides had caught anything that afternoon. Everybody wanted to know where we had caught such a monster fish, because most of them had been out all day and caught nothing remarkable. Joe just said, "Buoy 32" and would not share the location. That sheltered place in Pasagchak Bay was not used much for halibut fishing, and it was good to leave these large breeding fish to propagate the species for another year. I really had no intention of catching such a huge fish, though I suspect Joe knew exactly what was likely to be there and enjoyed surprising me. We gave about half of the vacuum packaged fish away to a local restaurant and put the rest into frozen storage, reserving some for dinner. Back at camp, we had halibut grilled over the fire with fresh bread and asparagus I had bought in town. The guys stayed up late telling tall tales, but I was sore in places I didn't know I had and went to bed to rest. Fresh halibut is delicious, and the adventure was priceless.

In the morning, I woke early to let Homer out for a short run before we headed back to Anchorage. The morning was misty and cool and as I stood waiting for the dog to finish running around, I heard muffled snorting and heavy footsteps nearby. I walked to the top of the ridge protecting the campsite from the road, and there stood a large herd of buffalo, about fifty beasts at least that belonged to the game hunting club farther out on the island. They came down from the mountain during the night to be safe in the brush from marauding bears and were trying to return to their grazing grounds further up the slope.

Well, there was a road crew arriving to begin paving this part of the road that headed out to the missile testing station at the end of the island. The buffalo were spread over the road and showed no inclination to depart. The workmen lounged against their trucks prepared to wait until the mist cleared and the buffalo wandered away. Some of the buffalo stood nearly as high as the truck cabs, so the men were in no hurry to have an encounter.

The foreman had other ideas. He was a bandy-legged, redheaded man with a really bad attitude. He just wanted the animals off his road so the crew could get to work. He stood in the middle of the road waving his hat and yelling at the buffalo. They were not impressed. The bulls lined up at the head of the column facing this irritant, with the cows and calves sheltered in the middle. The foreman was still yelling to no avail, and he jumped into a front loader and began to drive it toward the buffalo. They retreated a little way up the road creating a huge cloud of dust mixing with the rising fog. The foreman then declared it time to get to work and sauntered away.

But there was a rumble coming down the road and a thickening cloud of dust. Then the buffalo broke through the cloud and stampeded through the work site scattering workmen and sending equipment over the edge of the road. When the dust cleared, the buffalo had passed to their daytime grazing grounds beyond the work site, the men were picking themselves out of the bushes where they had jumped to safety, and the foreman was trying to get someone to help him right the payloader that had been driven over the ledge. Nobody was in a rush to help him. The buffalo had won the day.

I returned to Anchorage leaving Joe to meet his next group of fishing clients on Kodiak.

2.5 BEAR ENCOUNTERS

There are many horrific stories of bear encounters in the wilds of Alaska. Mine were limited but took place in settled areas where people had intruded upon the bear's habitat. I feared the bears and always took precautions when hiking, especially if I was alone in the woods, since I did not go armed. My dog, Homer, was my sentinel and I relied on his nose and quick sense of danger to alert me. On several occasions, he would bristle up his fur and growl, and I always made a hasty retreat.

Bear at Dawn

Soon after we settled into our log house on Oney Circle, I woke early to an unfamiliar sound at an unusual hour. Were the geese attacking the canoe? Was my husband snoring in a particularly resonant mode? I lifted my head from the pillow to scrutinize the rosy glow over the nearest mountain crest. A typical morning, but the snorting and loud crashing outside were clearly not the usual sounds. I rose to peer through the window.

The thrashing resolved into the frantic stomping of a mother moose standing in the pond. Right before my eyes, a shaggy form burst into view from the brush surrounding the pond. A small moose sprawled on the ground just beyond the dock at the end of the patio as a brown bear ripped it open. From the cover of the brush, I could hear the bleating of the other injured calf. The mother moose ran around the pond toward her calf, avoiding the bear.

He reared about and cut off her path reaching the calf first. I could not see the bear's actions, but the bleating ceased, the moose came through the water snorting and stomping at the bear as it dragged the second calf into the brush. She ran around the pond again, stopped and nuzzled the remains of her first calf and headed off in the direction of the bear. The bear returned to retrieve the first calf, which he dragged into the bushes at the edge of the pond just beyond our dock. He covered it with leaves and debris and went off up the hill.

Joe was watching from the porch, and he ventured outside when the bear went up the hill with the first calf. He was standing on the patio when the bear returned. It approached its kill, and Joe shouted and waved his arms to frighten it off; he called for me to contact the Fish and Game Department. I spent a frantic five minutes and finally got through to a trooper dispatcher who took my number and told me someone would call back. When the call came, an hour later, the trooper spoke to Joe who reported that Fish and Game would come around, close to 8:00 a.m. We spent the time having coffee and keeping watch over the small corpse of the calf. The mother moose was still in the area standing in the water and moving back and forth between her calf and the point of departure of the bear. She was clearly distressed and anxious—a very unpredictable state.

The F&G people finally showed up fully armed with tranquilizers and weapons, thermoses of coffee, and paperback novels to read while on "bear stakeout." They kept watch for another hour or so. In the meantime, people had called from the block above us, reporting that the bear had been seen dragging a moose calf into the woods of Prospect Heights entrance to Chugach State Park. F&G decided to remove the calf body from the edge of our pond and advised us to keep alert. The day passed without further incident. I went off to work, and Joe proceeded to his chores.

When I came home that evening, we had a philosophical discussion over dinner about the fortunes of bears and moose and the balance of these creatures with humanity in the way. I felt the need for a walk, and Homer was beside himself with restrained energy. Joe loaded the shotgun and we set out. We had a casual stroll through our well-worn path around and through our property. We were taking a turn around the house, inspecting the flower beds and looking at the signs of growth and moose damage.

As we came around the corner of the back porch, *there was the bear*! He was almost on the patio, only fifty feet away. He was following the scent of the calf F&G had dragged across the patio earlier. Now hungry, he had returned to the site of his earlier kill. As far as the bear was concerned, he left a perfectly good meal under cover of brush in a reasonable place. Now it had somehow migrated to a different place. Maybe some interloper took it. We stood looking at each other.

There he stood, as high as the clothesline pole and wide enough to block the view of the pond. I was breathless but calm in a surreal moment that seemed longer than it was. I will never forget looking at the bear eye to eye. He was a light brown bear, quite young. I could see the darker fringe on the edges of his ears; I could see the whiskers quiver as he sniffed and there was moisture around his mouth. The front paws hung down chest high with long claws. I remember thinking that the bear was probably hungry, and we were only people without significant tooth or claw. (Joe was armed with a shotgun, but we were very close, and one shot would not necessarily drop such a bear.) I don't know what the bear was thinking in that long, long instant, but he dropped to all fours and ran down the hill to the pond. He wandered around for quite a while sniffing the area, took a dip in the pond, charged the geese to no avail, and wandered off up the hill on the other side.

It was a rare opportunity to see a bear at close range. The evening light was golden and clear. The blue sky was reflected in the pond and the edge was doubled along with the geese and the bear. (The camera had jammed again, of course!) He was a graceful, magnificent animal, young and in prime condition. His fur was bronze with blond shoulders, weighing about 600 pounds. He looked hungry and impatient around the squawking waterfowl. The tracks he left in the soft mud measured ten inches from end to end with long claw marks. The mother moose was a young one, probably with her first season of calves. A more experienced mother moose would have protected one calf and sacrificed the other. This one ran around the pond and ran around the house in distraught frenzy, leaving the bear to its own mission of eating.

I wonder if the tragedy will end with the termination of the bear, or the moose, or the wildness? Absent humankind, the cycle of life and survival in the wild has no emotion. Life is unfair, by definition. Some die so that others may thrive. But part of us dies when the wildness is conquered. I am glad I had the rare opportunity to see such forces at close range. However, I will be happy if I never have another bear encounter. That once was close enough and the outcome lucky enough. I know I am no match for such a creature, no matter how armed. I would be too curious to be an effective shot myself. As far as I could tell, this particular bear lived and thrived in the area but had not been sighted on the residential property surrounding the Chugach Park. Soon, there would be many salmon in Rabbit Creek to keep bears well occupied and give the young moose time to grow up.

Termination Dust

In September 2005, we were living in Gridwood, south of Anchorage, in a rented house nestled in the cleft of the mountains, with a clear view of Turnagain Arm with its sparkling water and the snow-capped mountain across the way. "Termination Dust" of light snow on the mountaintops had come early this year, and the trees already covered the hills with gold from birch, aspen and alder, with the under story a bright crimson from the high bush cranberry. In this complex of six houses, there was a central shared bath house and laundry to which we could walk on a wooden walkway. On the banister, the neighbor's peacocks would often perch to wait for a handout of sunflower seeds. Their screaming gave an eerie and somewhat surreal tenor to the landscape. I was living there alone as Joe was still in Kodiak with clients fishing and hunting bear.

The house was heated with a wood stove, so my first task after coming home from work was to stoke the fire for the evening. I would find Homer and Kitty curled together on the cedar dog bed with Kitty snuggled against Homer facing the stove. Usually the fire had burned down to embers by the time I came home. The wood pile stretched several hundred feet along the path into the woods covered by a half roof. I kept a day's worth of wood stacked in the kitchen, but I needed to collect kindling from among the fallen branches in the woods. I set Homer lose for a run and stepped out onto the porch headed for the brush pile I knew was at the end of the path. But Homer was not bounding ahead of me down the trail as usual. He stood on the porch, hair raised with his nose sniffing into the wind, not moving at all. I looked off in that direction into the clearing of the cabin below mine. I did not see anything, but there was a distinct sound of crashing as if someone were tromping through the brush. I wondered whether some hiker had become lost and was trying to thrash through the brush from the road below.

As I stood with my hand on Homer's collar, the crashing turned into a large grizzly bear. It moved into the clearing and started sniffing in the fire pit, then stood over the empty garbage can, poked his head into it and pushed it over. Then, he turned his head and looked right at us. Homer began a low growl and was straining to run to the bear, a totally disastrous idea! The bear sniffed, turning his head from side to side, then he began to come up the trail toward us. I dragged Homer into the house, closed and barricaded the door and tried calling my neighbor. No answer. We were on our own with a really big bear coming up the path.

Images of bear breakins began to run through my head. I knew bears could rip doors down or tear windows out of their frames. Homer was by now barking his head off and jumping at the door. The bear was sniffing around the edges. I began to pound on the door and yell (as if *that* would do any good). I could hear the bear's claws scrape over the door surface, but so far it was holding. I kept pounding on the door, then I got the cast iron frying pan and began pounding on that with a hammer. It made a significant racket, and I figured I had a vanishingly small chance of clonking the bear over the head with it if he actually broke in. (Right. *That* will work!)

Finally, he gave up and shuffled off down the path through the woods and left. I told Homer he would have to wait for his run outside and I put down newspapers near the door for him. No way was I going back out there! I took the hatchet and skinned off shavings from one of the fire logs to make kindling, stoked up the fire and curled up with the kitty to watch the sun set behind the mountains. Hopefully the bear found something appropriate to eat, and hopefully he will be out of the area by morning when I have to brace for the walk to the bath house, then to the car a quarter mile down the path.

As I sat there trying to calm down, I realized there is a shotgun behind the front door, but I am sure I would be a hazard to myself and everyone around me if I tried to use it. What if I had gone out a few minutes sooner? I would have been at the end of the path, several hundred feet from the door, with Homer bounding off in the woods somewhere. I was so glad the bear did not approach the other entry way where the door is pretty shaky, and the freezer full of salmon was ready for plunder. This was a very close call and left me realizing again how vulnerable this place was. I was an intrusion to the natural habitat of bears and snowshoe hares, rabbits, and moose. This was not my place. I made my preparations and decided to go home, back to Pittsburgh.

PART III

HARNESSING EARTH'S HEALING POWER

"For the first time in the history of the world, every human being is now subjected to contact with dangerous chemicals, from the moment of conception until death."
Rachel Carson, *Silent Spring*, p. 12

Part Three

I came home to Pittsburgh in 2005 and took the position of executive director of the Rachel Carson Homestead Association, then the director of the Rachel Carson Institute at Chatham University. During this time, I found my life partner in Tom Jensen, who has shared my cancer journey and is on his own path of fighting cancer himself as I write this book. We live in a small borough to the east of Pittsburgh and enjoy the gardens, fish pond, and the company of Pasha Pussycat. I had the time and opportunity to reconnect with my family after living away for many years. The essays in Part Three share some of my reflections on coming home, growing older, and stepping into a more public leadership role.

3.1 HARNESSING EARTH'S HEALING POWER

When life seems unbearable, my instinct is to go home. My experience in Alaska ended in a triple trauma: the inability to get a flight out of Anchorage in time to say goodbye to my Father before he passed; being outcast from mainstream Alaska society over my position opposing the Trans-Alaska Pipeline tariff; and the escalation of my husband's paranoia causing concern for my safety to lethal levels. Although I had achieved high levels of professional success during my time in Anchorage, it came at the price of increasing isolation from my family and the sense of missing major milestones in the lives of my children and grandchildren. Seeing the little ones only once or twice a year was particularly frustrating and made me sad.

In September of 2005 as I was bracing for Joe's return from fishing on Kodiak Island, I resolved to file for divorce and leave town before he could return to Anchorage. I sent all my crucial papers and valuables to my brother in Pittsburgh, left my dog Homer in the care of friends, and fled to Pittsburgh with my cat Kitty Yum-Yum and two suitcases.

I settled in a house half the size of the one I left, across the street from my brother in Forest Hills Borough, a few miles to the East of Pittsburgh. The gardens were full of plants he had shared across the street with his neighbor over the years. I found Mother's sedum, peonies, German primrose, and iris, some of which came from my garden in Connecticut years ago. It was clearly meant to be mine. I planted a cutting from Michael's fig tree and returned to my roots.

I left the stress and chaos of the corporate world and took up the leadership of the Rachel Carson Homestead

Association to rebuild a legacy of living in harmony with Nature. I reveled in the luxury of giving voice and current relevance to Rachel Carson as my fulltime occupation. As organizer of the Carson Centennial Celebration in 2007 and the Silent Spring at 50 celebrations, I connected with Rachel Carson scholars all over the world, and had the opportunity to speak and write in her honor. I continue this activity in my retired life to this day.

Challenged by my own students, I ran for office and was elected to the borough council of Forest Hills Borough in 2016. I was reelected for a second term and have been honored to serve my community as we work on the transition to a center for innovation and preserve the forest landscape of our Tree City.

Connecting to the healing force of the earth sustained my battles with a kidney tumor in 2017 and breast cancer in 2018, made more poignant by sharing the experience with my daughter who was diagnosed with breast cancer at the same time. Here the path was quite hard because at this late stage of my life, I was nearly ready to give up. Dancing with the red devil of chemotherapy followed by bilateral radical mastectomy truly challenged my sense of self. The interface with the enormous cancer institution at Magee Women's Hospital in Pittsburgh was much more impersonal than my experience a decade earlier in Anchorage, where I was treated by people who were also personal friends. Having my daughter on a parallel path, but with a much more positive prognosis and outcome, tore me to pieces as I was helpless to be there for her in Virginia while I was in Pittsburgh.

In the recovery from this battle, defiance truly became my voice— a voice for the Living Earth that has sustained me my entire life. My battle with this fourth cancer crystallized my life purpose to transform our culture and civilization from one dependent on fossil extractive industries to one based on renewable and sustainable systems for energy, food, and materials. In so many ways, fighting to replace the fossil industries feels like a direct confrontation with the probable causes of my many tumors. The interlinked existential crises of global warming, global pollution, and global loss of biodiversity are manifest in the deteriorating quality of life and health of people everywhere, even in the richest countries and for the wealthiest people. I live on borrowed time but defiantly stand in the face of the challenges both to my own health and to the health of our society. I speak for the rights of Mother Earth and for all the children of the 21st century whose fate we shape each day by the decisions we make.

In the fifteen years since coming back to Pittsburgh, I have found my voice as a teacher, a political leader, and as a visionary for building a better future. In this turbulent time, the voice of empowerment and hope is critical to sustain the battle against the inertia of long habits that need to change. My two battles with cancer in this time have put a temporary damper on my physical capacity but sharpened my resolve. I have deleted the "urgent-unimportant" category of actions from my life and recaptured the joy of empowering others.

I watch my grandchildren, nieces, and nephews grow in talent and grace. I rejoice in the community of which I am an elected leader, and I take pride in the accomplishments of my students, many of whom are now leaders in their own communities. I have not allowed the invasion of kidney tumors or breast cancer to make me bitter or afraid because I know I live on borrowed time. Each day is a gift to be used to best purpose—to share joy and love with those around me, and to celebrate life as it comes. Each day I press the battle to preserve this living earth rejoicing in the triumph of the Spirit of Nature that lives in each of us for as long as we breathe.

I reflected on Rachel Carson's battle with cancer as I made decisions about how much of my own struggle to share. There is no hiding from the people I share daily life with—my partner, Tom, my son, and my family certainly know all the miseries. But to my public I hid under a veneer of being "fine." I wore a wig and careful makeup to give the illusion that I had eyelashes, eyebrows, and hair. I had to succumb to being driven because the neuropathy in my hands and feet made driving impossible since I could not distinguish between the gas and the brake. As much as possible, I presented a face of undaunted normalcy. I launched my first book, *Pathways to Our Sustainable Future*, and gave 35 speeches in 2018 while I was having chemotherapy and surgery and missed only one borough council meeting during the actual week of my surgery in October 2018.

I had a chance to speak at the Association for Environmental Studies and Sciences Annual meeting in 2018 as the immediate past William Freudenberg Award recipient where I addressed the power of joined voices. Only a few of my closest colleagues guessed that I was not well when I had to sit down afterwards with my feet up, sipping cold water. I became desperate to pass on the passion for active engagement, knowing my own time may be pitifully short. So, I got on my soap box and "preached." We closed with a round of "Who has the Power? *We have the Power!*", chanted back and forth three times. And I ended: "Yes, *we* have the power! Use it wisely and use it well. Blessed be!"

These essays written while I was undergoing treatment for cancer reflect the healing power of the ordinary things in Nature that surround us every day. The little gifts of the natural world carry the cycle of life across the stream of time.

3.2 THE POWER OF JOINED VOICES

During the time of both of my kidney cancer in 2017 and breast cancer in 2018, I was serving as an elected member of the Forest Hills Borough Council, located just east of Pittsburgh, and maintained my position in the Association of Environmental Studies and Sciences, for which I received the William Freudenburg Lifetime Achievement Award in 2017. These essays reflect some of my speeches and reflections on the strength and power that come from calling like-minded people together in common purpose.

Earth Day 2018 – A View of Hope from the Pit of Despair

A bright profusion of daffodils rims the pond. Young mourning doves explore the edge of the waterfall with their fuzzy plumage offering camouflage from the Coopers hawks soaring overhead. Blossoms and tree leaves swell in readiness to burst forth with the rich foliage of summer. I listen to the songs of the birds in their spring courtship calls and take comfort that the flow of the seasons continues. At the micro-level of a single back yard, the thrum of life pulses within the earth and gives me peace. So much of what gives life meaning is embedded in little things, priceless things like spring.

In the spring of 2018, my thoughts turned to Rachel Carson and her heroic battle to complete her book, *Silent Spring,* in the face of a metastatic invasion of breast cancer into her bones and lungs, causing every nerve ending to be wracked with the devastation of a disease that in her time was a death sentence. The one in eight women in America who face this same disease today have a much more favorable trajectory for survival.[5]

Rachel Carson's voice calling for precaution in the use of manmade materials that are biologically active had fallen on deaf ears. Even the protections for clean air and water and the toxic substances controls imposed by law failed to stem the flow of toxic releases. Now labeled as "burdensome regulations," even the minimum standards in place are under attack in favor of unfettered pollution to create short-term economic profits.[6] The myth that protecting the environment costs jobs is well entrenched and shows no sign of abatement.

I look at a trajectory forward from this year and see nightmare visions of rivers flowing black with coal waste, plastic suffocating the life of the oceans, air thickened by noxious emissions newly relieved of constraints. I think of the Pittsburgh of the late '50s when I was old enough to notice and complain of the sulfurous smell that suffused my world. Is it even remotely possible that this past will be the future my grandchildren know? I tremble in rage at even the possibility of such an outcome.

Rachel Carson's precautionary message, vilified in the industrial mainstream in America, has taken hold in the regulatory systems of other countries, especially Europe. In the EU, the burden of proof of chemical safety rests on the manufacturers who must demonstrate that products and their breakdown components pose no health or toxic danger to people or living things. It is not so in the U.S. Here, the industry meets minimum requirements, and whole categories of materials are "generally regarded as safe" without testing for health effects. The burden is on the consumer to prove that their illness was caused by exposure.

According to the Centers for Disease Control and Prevention biennial bioassay of the U.S. population, for example, the average American has more than 300 synthetic chemicals in his or her body, seventy-five of which are known mutagens or carcinogens.[7] Ninety-three percent of the adult population has Bis-Phenyl-A in their bodies, a known endocrine disruptor found in plastic container linings, thermal paper such as receipts, and plastics used for storing food.[8] Even babies are born pre-polluted, as documented by a study of cord blood in newborns that showed 237 synthetic chemicals present at birth, including carcinogenic and mutagenic compounds.[9] The wanton disregard for post-consumer fate of synthetic materials now forms a global chemical stew that surrounds all living things. The modern Age of Plastic has been a massive experiment on life without any controls.

Awareness of global pollution as an existential problem is growing across the world. It is impossible to ignore the millions of ocean creatures coming to land dead from consuming plastic debris floating in the ocean in great gyres created by the currents. Our habit of converting fossil raw material to trash as rapidly as possible with no plan for retrieving the waste creates millions and millions of pounds a year of synthetic material that does not break down into smaller molecules that can re-enter the cycle of life. This synthetic material is made from fossil resources, extracted with great damage to the living systems of the earth, then manufactured into materials for convenience. Three hundred million tons of plastic are produced every year, over half for single use items that become trash. More than eight million tons of plastic debris ends up in the ocean every year.[10] Modern living has hundreds of daily actions depending on plastics—structural components of buildings, vehicles, electronics, tools, instruments, and fibers. The problem of plastic pollution is complex and has evolved over at least fifty years. The solutions will require dedicated effort, but most critically, a force of will to change the process that leads to solutions.

It is a moral and ethical problem, not just a technology problem. The plastic pollution of the globe is the most serious unintended consequence of convenience combined with a failure to take responsibility for the waste produced at any level. Manufacturers have failed to take responsibility at the design stage to prevent toxicity and harm in the biological activity of the synthetic material they produce. Unless regulatory restrictions are imposed and enforced, there is no ethic for assuring safety in the products or their degradation by-products. Industry, especially in the U.S., screams about burdensome regulation and insists that restrictions limit profits and kill jobs. Producers of plastics, especially single-use consumer convenience products, take no responsibility for reclaiming or recapturing the waste.

There is no profit in recapturing the used materials, for it appears cheaper to make new plastics from fossil raw resources like petroleum and natural gas liquids. Retailers and advertisers promote ever more items for convenience, representing the single-use and throw-away concept as a convenience to the consumer. Cutlery, plastic cups, dishes, straws, food containers, takeout foods, packaging everything within packages then clad in shrink-wrap—the list is endless. Consumers take little responsibility for the waste created by all this "convenience." Americans recycle less than 5% of the plastic waste.

The ethic of taking responsibility for recycling plastic has evaporated along with the old-fashioned practice of returning beverage bottles for reuse. Soda, milk, beer, and water once came in bottles with a deposit which was refunded upon return. Glass bottles could be cleaned and reused five or more times before being recycled and re-formed for renewed use, a circular fate for the silica-based resource of glass. This practice is routine in Germany, where reuse of beverage bottles is standard.[11] They also recycle and reuse some plastic bottles with machines that shred the bottles at the point of sale for a deposit.

Solutions to the single-use plastic problem can begin immediately with citizens calling for responsible plastic policies.[12] *Refuse* single-use plastics like straws, shopping bags, water and soft drink bottles, cutlery, and food containers. *Reduce* the amount of disposable plastic in a conscious effort at the point of purchase. *Ask*, "Is there a reusable version of this product? Is the container recyclable? Is the packaging excessive? What becomes of this product when I am done with it?" *Plan ahead* to bring reusable shopping bags, reusable cutlery, cups, and water bottles. Bring reusable containers for takeout rather than Styrofoam or polystyrene take out boxes.

Reuse items that can be repurposed for creative applications from crafts to the selection of goods made from recycled materials, such as wrapping paper, carpet, flooring, and some furniture. Recycled plastic for 3-D printing and ocean-recovered plastic for product containers are two initiatives from industries developing more responsible global practices. *Recycle* responsibly. Know the requirements for recycling in your community. Sort appropriately; wash out food contamination; and avoid cross-contamination that will send the entire load to a landfill.

Raise your voice to demand better plastics–management policies. the local level, seek community action to have efficient recycling programs, compositing clean materials for community gardens while being careful to prevent plastic contamination. Stand up for state and federal rules that make product safety a priority to protect consumers. Call product manufacturers of the brands you use and demand a responsible waste recovery program. If nothing else, send the packaging back! In the UK, a group of consumers have been leaving excess single-use packaging at the store after check out. Call your representatives and senators to demand stronger regulations that protect consumers and the environment by reducing the production of single-use materials at the source.[13]

In 1970, the first Earth Day called millions of people to action. We filled the streets in droves, held teach–ins and demanded that lawmakers pay attention to the pollution of water, air, and land that was killing us and our children. Today, the approach of limiting the exposure by determining allowable levels of emissions has still resulted in

5.2 billion pounds per year of toxic releases into air, water, and land.[14] Today's technology has the capacity to go beyond the old adage that the "solution to pollution is dilution." We have managed to pollute the oceans globally, the air worldwide, entire watersheds, and acres and acres of farmland.

It is time to exercise the precautionary principle in full force. Design materials to be safe from the beginning—benign by design through green chemistry practices.[15] The culture of convenience based on consumer freedom to act without restraints and for industries to make decisions based exclusively on the economic profit as a driver leaves the priceless attributes of the living Earth exposed to wanton destruction. Freedom without responsibility and accountability for damage leads to chaos. The moral obligation to preserve the priceless life support system of the Earth must balance the economic drive of profits at any cost. We can live without plastic straws; we cannot live without fresh water, clean air, fertile ground, and the biodiversity of species that constitute the interconnected Web of Life, of which humans are but one part. On this Earth Day, take a walk through your neighborhood, and pick up all the trash around you. Notice how much plastic debris has become a normal part of the landscape and resolve to be part of the solution.

Hope for Survival

I awakened this morning to the chorus of bird song unmarred by the weekday racket of buses and cars from the school next door. As I listened to pick out the individual voices of familiar friends, I felt once more the surge of joy in being alive in such a beautiful, wonderful place and time. On this May morning, the pink phase of my garden was in full evidence. The redbud tree stood with cascades of lovely flowers and the magnolia blossoms accent the middle space between the paler blossoms of the crabapple and the serviceberry. The sun fell bright through the branches of the great pin oaks with their leaves still furled and unfolding slowly. This neighborhood of soaring trees and diverse wildlife habitat offers hope that we can live in harmony with Nature.

Yet, as I stepped outside to have coffee on the patio, the sulfurous smell of the coke works beyond the hill assaulted my senses again. The continuing coal use lingering from the past still pollutes the air and compromises the health of all who live here. I think about the people who live in the areas closest to the power plants, the coke plants, the chemical factories, and the plume of highway fumes from cars, trucks, barges and trains. The trappings of our modern life press upon the natural world on all sides. It is not enough that I live in a little place surrounded by trees laden with flowers and exploding with life. It is necessary that we have living communities as the normal base line of life in America, not the exception for those who have a plot of land and a house. It is essential that we build a way forward that promises a better future for all our children and the generations to follow.

As I listened to the conversations of the robins, the calls of the cardinal, and the trill of the goldfinches, I thought again of Rachel Carson. She who sought solace at the seashore never forgot the interconnectedness of all of life on earth. The daily headlines document the deterioration of our world, its oceans, forests, streams, islands, prairies, and wetlands—all are under attack.[16] Seventy-five percent of Americans consider the environment a top priority, but the positions are deeply divided along partisan lines, and even more steeply between young and older generations.[17]

Though preserving environmental quality seems important to people, this issue does not rank high in the political discourse for major elections. Fossil industry interests still control the narrative of public debate, asserting their entitled position that places rights to the extraction of fossil reserves above the rights for living systems to exist. People do not connect preserving forests, rivers, wetlands, grasslands, estuaries, prairies, and wildlife with their own personal quality of life. Only 32% of millennials place environmental issues as a top priority, and fewer than one in four Americans practice lifestyle choices with environmental values in mind. Without broad support for environmental conservation and protection, especially among millennial-generation emerging leaders, there is little hope of changing the tragic trajectory of our culture.

Among people who are concerned or seriously alarmed about climate change or pollution, very few act. Often labeled and marginalized as "tree huggers" or other more insulting terms, environmentalists are shunned as alarmists or unrealistic idealists. Most people who are concerned consider the issues of climate change and global pollution to be beyond the influence of individuals. People are waiting for some luminous visionary leader to "save us." Many people pretend these issues will go away. They hope technology will emerge to correct the problem, and some even think it is God's role to protect us, so therefore humans should not interfere.

I wonder still at the battle waging within my body as the technology of chemotherapy goes 26 rounds with my tumor. I wonder whether I will survive, even if the tumor does not. Sometimes, I see the pollution around me and think of it mirrored in my cells. The "micro-inclusions" in my own lungs were detected in one of the chest X-rays I endured. The nurse shocked me when she asked how long I had been a smoker. I have never smoked! She shrugged and said, "Oh, you have Pittsburgh lungs. Most people who grew up here in the '40s and '50s have little pockets of particulates in their lungs from breathing all the smoke from the mills." This has me wondering what other baggage I carry from those early days.

And I wonder about the burden imposed daily on today's children from the coke works that have been running since 1837, grandfathered from air pollution controls, as well as newer pollution sources that send volatile organic compounds into the air and heavily saline brine full of radioactive and toxic materials that come from fracking into our water. How many of them will become victims of asthma and tumors or neurological disorders? I contemplate with dismay my own contribution to the inundation of plastic waste that has marked our time.

Each chemotherapy treatment generates a huge amount of plastic toxic waste, not only the intravenous infusion tubing but the packaging it is delivered in, the protective gowns, gloves and masks the nurses wear, and the plastic disposal bag into which everything is thrust when finished. Even the plastic port in my chest will end up in a toxic materials landfill. My own bodily wastes are toxic too for at least 24 hours, so I am adding to the burden of contamination to the biosphere. Is this treatment worth it? Isn't there any other way to address this tumor than by poisoning me and everything around me?

As I sit quietly, the resilience of the birds that grace my gardens amazes me. I know they are fewer in both number and kinds than a decade ago. Their habitat has been weakened by the continued incursion of development, and by the constant barrage of pollution. But their continued presence as part of my world allows me a hopeful perspective. This is worth preserving, and life is worth fighting for. The cardinal and Carolina wren exchange arias of joy. I can add my voice for change to theirs and make some small restitution for my share of their burden. During the summer of my chemotherapy, I gave a number of presentations as part of my book launch focused on moving from awareness to action. I also began working with a coalition of friends and associates to develop a movement to empower communities. Here is one of the distillations of this action, written on my birthday May 20, 2018.

Moving from Awareness to Action

It took hundreds of millions of years to produce the life that now inhabits the earth — eons of time in which that developing and evolving and diversifying

*life reached a state of adjustment and balance with
its surroundings. Given time – time not in years
but in millennia – life adjusts, and a balance has
been reached. For time is the essential ingredient;
but in the modern world there is no time. Rachel
Carson.* [18]

Daily headlines document the devolution of our
environmental protections, even as the conditions of climate and pollution grow worse.[19] A numbing effect sets
in; beyond disbelief, a paralysis of will sends people into
a shocked retreat. We pretend that some visionary leader
will step in to save us. Or that a yet undiscovered technology will emerge to reverse the effects of global warming
and global plastic pollution. We pretend it will all be fine
and try to go on with our lives while the basic life support
system of our earth is torn to shreds. It is the children who
are outraged, who bring suit and scream for justice.[20] It
is the Native American defenders of water and land who
rise up with their lives on the line to protest and object.[21]

When rules protecting endangered species, drinking water, farm workers and children are dismantled in
the name of immediate profits, or the lure of jobs, where
is the outrage against the harm? Against the injustice? In
nine states, laws are under consideration that would make
protesting energy infrastructure a criminal act, subject
to prison as domestic terrorism.[22] Where is the outrage
against the basic violation of First Amendment rights to
free speech and assembly? When did cruelty become a
value that makes America "great"?

No visionary leader is going to come forth to save
us. *We* must take responsibility to object directly to those
in office at all levels who are making these decisions. *We*
must take action ourselves in our daily lives. There is no
way to generate the necessary uprising of protest against
the outrageous actions of this administration and those
complicit by silence without *each one of us* standing up
and declaring *enough*! There is a better choice for a way

forward. We have better options for our economy, for our way of life, for our children's future. We do not need to destroy the Earth to have a thriving civilization. Indeed, we must preserve and restore the living systems of the Earth if we are to survive at all.

On this day, my 72nd birthday, I call on all my colleagues and friends, collaborators and associates to *stand up! Speak up!* End the complacent silence that gives tacit permission for the destruction of our world to continue. We must exercise our obligations as citizens, as caring human beings, as children of Mother Earth to preserve the life support system of our planet. I urge a call to action as a manifesto for the environment.[23] The rationale for this call to action rests on the following facts:

- The constitution of the Commonwealth of Pennsylvania States *"The people have a right to clean air, pure water, and to the preservation of the natural, scenic, historic and esthetic values of the environment. Pennsylvania's public natural resources are the common property of all the people, including generations yet to come. As trustee of these resources, the Commonwealth shall conserve and maintain them for the benefit of all the people."* Article 1, Section 27.[24]

- All forms of exploitation, abuse and contamination have caused great destruction, degradation, and disruption of Mother Earth, putting life as we know it today at risk through phenomena such as climate change.[25]

- Communities and people of color have been disproportionally affected by the environmental, health, social, and cultural effects of energy and resource exploitation and development.[26]

- Burning fossil fuels, the principal cause

of global warming, compromises the life support system of all oxygen-breathing, fresh-water-dependent organisms, including humans, while global pollution from man-made chemicals, especially those with endocrine disrupting properties, threaten the health of creatures throughout the world.[27]

- The health and well-being of people and especially children are significantly degraded.[28]

- One in twelve Americans suffer from asthma.[29]

- In 2018, an estimated 1,735,350 new cases of cancer will be diagnosed in the United States and 609,640 people will die from the disease.[30]

- Newborn babies have more than 200 synthetic chemicals in their blood, 75 of which are known to cause mutations and cancers.[31]

- Sperm counts have declined by 50% to 60% in the last 40 years in America and other Western countries.[32]

It is critical to seek and support people in office at all levels who support the following positions:

1. To protect, restore, and preserve for future generations the fresh water, clean air, fertile ground, and biodiversity of species of Pennsylvania, the United States, and the world.

2. To promote urban and rural ecological policies to clean up and rebuild our cities and rural areas honoring the cultural heritage of all our communities.

3. To support investment in renewable energy systems and regenerative agriculture while training workers to pursue careers in these fields.

4. To oppose destructive practices such as slick water hydraulic fracturing for oil and gas, destructive coal mining practices, and wanton pollution of water, air, and land.

5. To promote non-toxic manufacturing with an economy designed to reclaim and reuse materials, such as recycling of glass, plastic, paper, and metals, and to limit or eliminate single-use plastics.

6. To promote policies based on mutual respect and justice for all peoples, free from any discrimination or bias.

Only through the joined voices of all people who care about the future, about our children, and about the quality of life for all living things can we overcome the culture of greed that has evolved in America. The only value that matters in the current decision-making process is the dollar, the short-term economic benefit to interests vested in the existing political power structure. It is time to reassert the values of social equity, care, and concern for the elderly, ill, weak, and the children of our country.

Many people in past generations—especially unionized workers—have fought for the protections put into place over the past 100 years. Their efforts changed the laws to protect worker health and safety, cleaned up the air and the water, established wage and labor protections so that life expectancy increased, worker safety and health become a priority, and broadly shared prosperity was accomplished alongside of real progress in cleaning up the environment. Those successful battles also made it possible for people to enjoy our national and state parks, not only because these areas were protected but also because of the negotiated rights of workers to have time away from work available for themselves and their families.[33]

Everyone alive today has received the legacy of the struggles of the activists who came before us. What has been so hard won with blood, sweat and tears can be lost through indifference and complacency. It is time to reclaim and rebuild a public education system that prepares all Americans to respond to a changing future. It is time to have healthy people and a healthy environment as a right for everyone. It is time to reclaim America as a land of hope, empowerment, and caring communities instead of a place of ignorance, deprivation, and fear.

We must *each* stand up for what is true and right, with courage, determination, and passion. It is not enough to grumble to each other, to wring our hands and complain. It is time to act boldly. We do not want to see hard-won environmental protections rolled back to 1985 or earlier. We do not want to see worker and child labor laws weakened or rescinded. We do not want to have education become a privilege of the elite. We do not want toxic emissions to air, water, and land to become even more pervasive. A true democracy depends absolutely on an informed and engaged citizenry, on freedom of speech and of the press. We must stand up for our America, or we will be inhabitants of a despoiled and tortured land, her wealth squandered, her beauty plundered, her heart broken. To accept tyranny in silence is to become compliant in the slow murder of our culture.

We all live but a moment in the span of time. It is our privilege and our duty to make the most of our time on this Earth. We cannot know how many hours we have to spend, but we can commit to celebrate every opportunity for joy. We can weave ourselves into the tapestry of our time and immerse ourselves into the life-giving force of the living Earth. We can stand in defiance of the sadness, pain and evil that rises around us. We can be a beacon for those who follow, triumphant in living in harmony with Nature.

I will fight for clean air, fresh water, and fertile lands, and to preserve the beauty and wonderful intricacy of Nature to my last breath. Join me, for now, and for the unborn children of the 21st century whose fate we shape by our actions—or our silence. Joined voices of the People will prevail over tyranny and greed.

3.3 GIFTS OF
THE HEALING TREES

I have the privilege of living in the shelter of two great pin oaks, measured at 118 and 120 years old. These were my elders, sources of wisdom and strength during the long periods of my illness. These elder trees became my touchstone for healing as I reflected on all that had passed under their branches, all they had survived to be standing tall and wise embracing and arching over my small plot of land. Dwarfed by their stature, I am awed at their resilience and fortitude.

The Wisdom of the Great Pin Oaks

A soft rain was falling with a patter on the roof and the swish of water and wind in the heavy leafed trees. The chorus of morning bird songs defied the distant roar of early traffic— robins, cardinals, wrens, song sparrows, a Carolina wren. The air blew from the east this morning, without the usual traces of sulfur from the coke works, cool enough to allow open windows. As I sat in the garden, I let the sense of this space seep into my consciousness. Here the trees tower over the house and the gardens flow from one to another connecting with the woods. I seek inspiration from these elders of the neighborhood, the hundred-year-old pin oaks that preside over my small plot with their canopy spreading edge to edge of this space. Under the grass and flowerbeds, their roots run deep to hold the ground and to take counsel each to each through the myriad micro-connections that tell them the state of their world.

I left a small sapling growing in an auspicious place between an oak and a maple to take succession for the next hundred years. Its maturation can proceed at the foot of its elders, linked root to root, absorbing the wisdom of ages on the ways of trees and people. What conclusions do they draw from the air samples absorbed through their stomata? Surely the lack of heavy smoke of past days is a relief, but as they detect a shift in the carbon dioxide concentration in the air, how will they respond? What do they think when they scent the presence of strange compounds of man's creation?

Will these trees, sprouted before I was born, stand here long after I have departed this earth? Will they grow for decades, providing shade, food, and shelter and purifying the air for the creatures that share this space? Or will they fall to a super storm, drought, or disease, finally releasing their stored carbon through the slow oxidation of decomposition? We cannot know these things. We can only sit and commune together in this early morning while the soft rain falls, and the birds greet the rising sun.

In the afternoon, I rested on my afghan on the unmown "lawn" and allowed myself to sink into the ground as I recovered from this week's chemotherapy treatment. The blooming clover was alive with honeybees, most likely from my neighbor Natasha's busy hives. Other insects also hovered and flit in this blooming mini-meadow. An improbably bright blue dragonfly zoomed by. A pair of tiger swallowtail butterflies flirted in the sky above me. The robin parents took turns going back and forth to feed their hungry chicks in the mock orange bush. I heard the piping of the chickadee nestlings in the birdhouse behind the pond.

I could feel the warm earth below me and sense the life below the surface. Here earthworms and nematodes ply their craft. Unseen millions of soil microbes break down debris to the nutrients drawn in by the thick mesh of tree roots below me. I see them reflected in the interlaced branches of the oak canopy above me. Sunlight filtered through the thick leafy cover to send dappled light down to my resting place. A few small caterpillars drift on their spun threads, and some fall prey to the birds that work their way through the tree canopy calling chip notes to each other assuring that all is well.

I closed my eyes and felt the embrace the elder pin oaks extend to me. These great trees form the connection between earth and sky, each one an engine of carbon sequestration as their leaves turn sunlight into sugars and starches and cellulose. The miracle of photosynthesis happens on a grand scale in this summer of rain and warm weather. The air so close to the ground holds an earthy fragrance laced with clover blossoms and the star gazer lilies in the nearby garden border. My senses are dulled by the chemicals coursing through my body in the attempt to stifle the tumor growing in my breast. But saturating myself in all of this pulsing life lends me a sense of calm and confidence in the enduring resilience of Nature. I listen to the soft gurgle of the pond fountain and receive the gift of healing sleep.

The Healing Trees

I lie in the grass on this late June afternoon, relaxing after a day of making bread and hanging laundry on the line to dry in the soft summer air. As I let myself sink into the sense of being one with the earth, I remember the many afternoons I would lie here on my afghan after coming home from chemo treatments in the long, sad summer of 2018. Now in the pandemic summer of 2020, the world has shrunk to an immediate family "bubble." We stay in our pods conducting business on Zoom, spending time on the internet, trying to network and participate in the usual election year routines of canvassing, phone banking, and postcard writing without the physical closeness of the community of like-minded people who support each other through this kind of activity. This COVID-19 summer has reshaped what community means—when staying away from each other is a form of care and social-distanced interaction defines friendship.

Here on eye level with the grass, I watch the honeybees hum from blossom to blossom in the clover filled un-mowed lawn. As my gaze drifts over to the herbaceous border around the pond, I find myself eye to eye with a green-mouthed bullfrog grown from one of the tadpoles my grandson gave me for a birthday present. It has lived in my pond for three years now and has a clouded right eye but seems fat and healthy. Under the leaves of the hydrangea bush, I spot a resting firefly. Later this evening, the garden will be full of these magical creatures as they perform their flickering mating rituals.

A column of ants streams away toward the stone wall from a piece of watermelon rind that was strewn into the edge of the garden when my young nieces and nephews visited earlier this week. Overhead, the interlacing branches of the century-old pin oaks shelter birds, provide sanctuary to myriad insects and provide a general highway for the marauding squirrels. Deep under the ground, the fungi and mycelia send chemical messages from tree to tree monitoring the moisture, nutrients and state of the world.

As I lie there allowing myself to sink into the ground, I close my eyes and imagine the layers of interconnectedness that cradle me above and below, connecting the soil to the sky. I breathe in the scent of blooming flowers nearby and feel my own place in this web of life. This is the intimate personal community of living things that share my daily life. Most often unseen, there are thousands of creatures going about their business as pollinators, foragers, detrital consumers, aerators of soil, exhalers of oxygen. This profusion of life powered by sunlight thrives all around us, usually without notice. This is the fine fiber of the web of life, minute and intricately woven down to the molecular biochemistry that turns sunlight into sugar and starch through the miracle of photosynthesis.

As I contemplate the more than 300,000 lost to the COVID-19 virus so far this year in America alone, I realize that the human tragedy has no effect on this space of living earth. The ecosystems we depend on as our life support system do not need humans to operate, but humans have the power to disrupt and destroy them.

Humans are not an essential part of the living Earth but have come to dominate over three quarters of the planet, for the most part in destructive ways.

Millions of acres of once bountiful mountains are laid bare to the bedrock to extract thin coal seams. Once free-flowing streams are choked and filled with the spoil from removing mountaintops. Coral reefs once teeming with fish stand bleached in acres of underwater ghost reefs. The permafrost of the northern tundra melts and softens, releasing great gasps of methane. The polar ice caps melt, inexorably raising the sea level and altering the flow of the great belts of ocean currents that cycle nutrients through the ocean. All around the world, people continue to burn fossil fuels for power, for transportation, heating, and cooling houses, and manufacturing goods that rapidly become trash. The pandemic of COVID-19 shakes our foundations, a symptom of the broken systems that provide checks and balances. This madness must end! The earth can heal, but we must help.

The resilience of the living earth rests on the laws of nature, which are not negotiable. People must change to align with them. We must transform our extractive and destructive civilization to one that rests on regeneration and preservation. This transformation is not a technology problem. Rather it is an ethical challenge to restore and preserve the life support system of our planet for our children.

Every time I see a flower bloom from the crack in a sidewalk or find a small tree in the space at the edge of a building, I know that life will triumph over even the most insensitive human folly. The Earth provides all that we need to survive and thrive if we only open ourselves to trust that we are part of the living world. We can live in harmony with Nature and find a shared prosperity where life will flourish.

3.4 WINTER REFLECTIONS: THE BEGINNING AND THE END

February 2018 Diagnosis

Fat snowflakes swirl and cling to branches and twigs coated in ice as the "wintery mix" storm persists into a second day. The bird feeding station attracts a varied flock, including three banded song sparrows and cardinals, and the unlikely pair of flickers pecking off the remaining poison ivy berries that cover the locust tree across the street. If the sun comes out, it will look like a diamond fairy land, but in this grey, cloud-covered afternoon, the chill and starkness of the winter suffocate any sense of gladness. In such a day, receiving bad news seems almost appropriate. The beautiful world I love shares my impending despair.

As I have been struggling to maintain a positive attitude in the face of daily assaults on social and environmental protections hard won over decades, I find myself facing a personal crisis of a serious negative diagnosis. Once more, I face an invasive cancer with an uncertain outcome. This is the fourth tumor—and maybe the last. It seems impossible once again to face the prospect of surgery, the debilitation of chemotherapy or radiation, and the agonizing reconstruction of a life from the shreds. Gearing up my emotional reserves to help my daughter through a breast cancer diagnosis has helped me a bit to prepare for my own challenge. At least we will be going through this together, at overlapping times, but separated by distance from tangible daily support. I am adamant that I will defer my own treatment to go to her for support while she is in the hospital, and I will preserve the most important commitments and opportunities I have before me.

As I watch the blowing snow pass the windows, I realize that the cycles of the seasons enforce a time of rest and withdrawal from the pace of growth. Deep under the ground, the mighty oaks continue to connect to each other through their interwoven roots. The praying mantis egg cases bide their time, sheltering the hundreds of little ones that will hatch out in the warm spring days. The swollen buds of the azalea and rhododendron lie waiting to burst forth in flower, soon—but not too soon! Under the ground, the bulbs planted last October will send forth their shoots and grace the gardens with daffodils, hyacinth, crocus, and snowdrops.

The snow now covers all the features of the landscape with a thick blanket. Gone is the mess of the leaf-mulched border gardens. The untidy brush pile is buried in a drift, with little spaces open for the sparrows to perch away from the storm. In the morning, I will look for the stories in the snow, perhaps the wing-print of the owl snatching a mole from its snow tunnel, perhaps the trail of the rabbit from its burrow under the brush will thread through the low bushes that still yield soft bark. I know the squirrels will be foraging among their buried troves of acorns and purloined sunflower seeds. I watch the little Carolina chickadee snatch a seed from the feeder and take it to a nearby branch to crack open and devour. I marvel at the resilience of these little creatures and their adaptation to the cold and the storm.

Winter Solstice 2019

Days and days of rain and clouded skies succeed an early snowfall. The frozen ground deflects water into drains and stream beds winding their way down the hillsides of this tree-covered neighborhood. Here most of the houses are smaller than the overarching canopy and branches interlace across property boundaries. Many neighbors also interconnect at this season, for parties, for informal invitations for coffee or greetings. The interconnectedness does not end with the dreary weather. The evening is brightened with holiday lights in windows and in lighted garden and house displays.

For the deciduous trees and the plants of this biome, this is the time for rest. The sap retreats to the depths of the ground, secure in holding to the Earth until the warmth of spring signals the time to rise and fill the budding leaves with life-giving nutrients. If the leaf fall of the previous season rests on the ground to cover and protect and later to decay and return the elements back to the ground, the cycle is complete, enriching the soil with each year.

In these short days and long evenings of winter, there is time for reflection, for writing, and laying down thread in a long-delayed quilt. For me, it is a miracle to see this winter after a long year of battle with breast cancer. Thankfully, the scourge of this disease has been set at bay one more time, making these winter days free of pain and so much more precious, free of drugs, free of exhaustion. I think of all the afternoons spent lying on the garden settee or on the grass, gazing through the interlaced branches of the pin oak elders above me, and feeling the thrum of life running from the ground to the utmost edges of their spreading leaves. Trees full of life force, supporting endless numbers of insects, visited by birds and squirrels and chipmunks, included me in their domain. I opened my heart to their healing energy and felt myself a part of this miracle of the living Earth. Healing is a state of mind. The technology of medicines and surgery deal with the mutiny of cancer cell growth, but the battle to overcome and to survive takes place in the mental space that holds the force of will to live.

I realize that these elder specimens have witnessed great changes in the world around them. They have stood here for more than a century, first as seedlings when this area was a dairy farm, growing up amid the smoke-filled air and volatile emissions at the height of the steel mills operating over the hill along the Monongahela River.

They witnessed the change from farmland to houses, fortunate that trees were valued in the landscape and were not bulldozed into flat acres when the houses were built.

Now, many are experiencing with us the strange weather patterns of a warming planet, driven by the very emissions that stunted their growth in the decades of the Industrial Revolution. Some have fallen to storms and high winds. Some have fallen to drought and strangulation from invasive ivy. Others fell from boring insect invaders. These two pin oaks stand as sentinels, guardians at the top of the hill, giving testament to the resilience and stability of the living Earth.

With great humility, I see them now as mentors and models of a way forward. There is no path to a sustainable future that does not include protection for the natural world. They store the wisdom of the ages in their collective interconnectedness. It is only humans who are cut apart from the life force of the Earth. We live under the delusion that technology is our salvation, that human knowledge can outwit the changes we have wrought upon ourselves. It is not so. It is only by embracing the force of the natural world that humans will survive and thrive.

The harmony of Nature's laws has evolved over many millions of years, fine-tuned to the ways each part of the biome affects another, each small piece contributing to the whole. We see daily reports of how the insects are declining worldwide, how coral reefs have bleached to dead skeletons. With increasing numbness, we hear of the extinction of creatures and plants—of whole ecosystems. We are seeing the harbingers of our own fate. The preservation of the living Earth is our only hope. We must recognize that humans are only one part of the natural world, intimately dependent upon the health of the living things around us. We thrive too when the butterflies and birds are healthy. We flourish when the songs of frogs fill the summer night.

It is my hope for the coming years that I can resume my quest for our communities and our nation to transition to a civilization living in harmony with Nature. I am thankful for the chance to be a part of this great web of life for one more year. I dedicate each day to living in harmony with Nature.

3.5 SWEET RAIN AND BITTER TEARS

March 28, 2022

The garden is full of birds again. The construction across the street that has marked each morning for the past two years with the piercing back-up beeping of construction vehicles has ended for now. The border hedge once again attracts the cardinals, tufted titmouse, and robins. The dawn chorus rings robust once more as the birds welcome a new day. I wake listening to the cadence of the cardinal duet, the robins cheerful chirping, and the intermittent call of the Carolina chickadee. So much life calling to me to cheer up and come outside.

But I lie a few moments more with Tom's raspy breaths next to me and just let the tears soak into my pillow. It has been a week of total misery. On Monday, Tom was treated for 23 more tumors in his brain, had a fifth chemo treatment on Wednesday, and on Thursday what I thought was going to be a routine echocardiogram and stress test landed me in the hospital for a cardiac catheterization. At the same time, my dear daughter was having surgery to remove a recurring malignant lump in her reconstructed breast. I have been overcome with tears, bitter with anger and frustration all week. I take my morning meditation into the garden to explore this new dimension of my sadness.

Early spring in Pittsburgh brings fickle weather, sometimes soft and filled with blooming flowers then abruptly freezing; daffodils drooped to the ground; nubs of green where the deer had munched the tulips. But when the sun comes back, the forsythia sparkle bright yellow along the borders, white and yellow daffodils raise their heads and welcome the early honey bees fliting about, paying special attention to the carpet of tiny wild hyacinth that make a blue haze in the undergrowth. They have propagated freely in the last few years and now venture even into the spaces of the lawn.

I move the leaf cover from some of the borders and there discover primrose emerging, their buds still closed, waiting for a bit more warmth. The new dandelions and red sorrel in the garden are ready for microgreen additions to salad along with the wild green chives that come up everywhere. I think the lavender and parsley and sage have survived the winter and will soon contribute their fragrance to the kitchen.

This is a foggy morning, and wisps of cloud drift up to dissipate in a swirl at the treetops. As I walk around the garden, then the block, noticing everywhere signs of life reemerging, my spirit calms. I marvel once again at the miracle of spring. The redbud and pear trees wait for the right sunny day to burst into flower. The stalks of last year's borders tip in the wind, sheltering the pupae and overwintering egg cases of praying mantis, beetles, and moths. As I walk in the garden path, the earth exudes that characteristic warm fragrance that gives promise of the rich vegetation that will soon break forth.

Already, the small points of lily of the valley and hosta pierce the leaf litter. Fern fronds rest furled and ready to expand into feathery plumes along the stone wall. I contemplate where to put new plantings of Helibore and make plans to put in more Scottish bells. The trout lilies carpet a section of the upper garden, and I lean over to see their yellow centers hanging down among the mottled leaves. They are too fragile to pick and quite ephemeral in their time of showy bloom, so more precious.

As I walk around the perimeter of my garden, I startle the female cardinal building a nest in the mock orange bush next to the garden door from the garage. She sits on the fence surrounding the garden and gives her alarm call, soon joined by her mate, and I back away to leave them in peace with their construction. My direct route now cut off, I wander through the garden again to reach the front door, and as I pass under the oak trees, a soft rain begins to fall. It is almost as if the condensed fog just decided to fall down instead of swirl up. Everything is covered with warm mist. I tip up my face to the clouds to look through the heavy branches of the great oaks. The rain mixes with my tears as they wet my cheeks. I take off my glasses and feel the sweet, warm rain on my face. I know that surely the spring follows the winter and I will feel joy and peace once again after I pass through this time of pain and sorrow.

PART IV

LIVE IN HARMONY
WITH NATURE

I pledge myself to preserve and protect America's fertile soils, her mighty forests and rivers, her wildlife and minerals, for on these her greatness was established, and her strength depends.
Rachel Carson's Conservation Pledge

Part Four

I have been teaching ethics and public policy at the University of Pittsburgh and Chatham University since 2007 but retired from that position in 2012 and wrote my first book, *Pathways to Our Sustainable Future*, based on the lectures from my class. As an elected official, I found many more occasions to speak formally in a leadership position to organizations and to other communities. I was fortunate to be on the team that built a new borough building for Forest Hills—a passive solar designed, net zero energy building. I received several awards and used my platform to spread the vision of a better future. People will not move to something they cannot visualize and see as an improvement on their current state. This section has some of my speeches and presentations from my post-cancer recovery period.

4.1 LIVE IN HARMONY
WITH NATURE

Facing a life-threatening illness forces one to focus on what is truly important. Every person meets such crisis-induced inflection points differently. As I have coped with four different challenges to my health over a span of twenty years, I have made decisions to live each day to the fullest, with purpose and intention. One day at a time, I rejoice in the wonder and beauty demonstrated everywhere through the gifts of the living earth. I seek ways to use my voice and my personal power to move the world around me to a more sustainable and resilient place through local political action as an elected official, through regional collaborations with like-minded colleagues, and through writing and focused contributions to national and international efforts. But all of this fades away in the face of a truly life-threatening reality.

When the days ahead are numbered to a few hundred at best, it is the relationships, the personal connections with a caring community of family and friends, that make the difference. All the time spent on causes and external concerns diminishes in significance compared to spending an hour in lucid conversation with a dear loved one. Memories of shared joys lift the pall of pain and fear. Simple pleasures enhance the sense of being connected and not alone in the darkest of times. Just holding hands and smiling through internal tears and broken-hearted grief give comfort.

When you become a patient in the institutionalized medical system, with a "care team" and a treatment plan, personal connections become critical. Who is the person who can understand the jargon and translate information into meaningful communication? Who can see through the doctor's shield that comes down over demeanor when the diagnosis is a condition without cure, just a "management plan"? In this situation, it is the inner strength of each person that sustains life with dignity and quality for as long as possible.

The ability to connect with the healing power of the living Earth makes an enormous difference in the experience of coping with a critical illness. Whether the condition will abate sufficiently to allow many years of living, or whether it is so acute that there are few options for prolonged life, living each day becomes either a gift or a burden, depending on the attitude and mental and spiritual support system of each person.

I remember my grandfather Pop in his late years when he was living with my parents. His Parkinson's disease had advanced too far for him to live alone, and he resented his loss of independence. He would sit on the bench in the patio under the pear tree and talk to Nona who had died years before. He would say, "Well, Pasqualine, the Lord forgot me again today. I'm still here, and you're with Him. How long must I wait to be free of this world?" And yet, when I came to visit with my two small children, his great-grandchildren, he would smile and sing them the same Italian songs he sang to me as a child. He would give them a ride on his foot, holding their little hands and bouncing them up and down. For those moments, he was alive and sharing experiences with another generation. They have not forgotten him, and the memories have crossed through generations.

These essays reflect my ongoing work as an advocate for the living Earth and a witness to the resilience and goodness that can come from people standing together for the sake of humanity. Rachel Carson said, "Those of us who know have the obligation to speak." I have taken her admonishment to heart carrying my own sense of purpose into the public arena of advocating for the living Earth as long as I have breath. Here are some essays that represent the major themes of my public lectures. You can see my ongoing work on my web site https://patriciademarco.com in my blog posts "Transformations for a Sustainable Future."

4.2 "WATER IS LIFE"

Lessons from the Standing Rock Sioux
Pipeline Protests

Fresh water lies at the nexus of two existential crises of our time—global warming and global pollution. Two mutually exclusive visions for the future played out on the streets of Pittsburgh in October 2018. The Shale Insight Conference at the David Lawrence Convention Center gathered gas and petrochemical industry corporations, workers, and supporters to share development plans and hear President Trump present his vision for this area as "the energy hub of America." Throughout the day, three protests organized by Bend the Arc, the Indigenous People Water Protectors, and the Women's Climate March demonstrated against the petrochemical build-out plans calling for protection of the water. A week later, Mayor Peduto speaking at the P4 Climate Summit, decried the petrochemical build-out in Western Pennsylvania as a backwards looking development, painting a vision for a more resilient and sustainable future for the region. The two messages define the great divide that is pulling America apart, but within the controversy, elements of common ground have the potential to unite all of us in common purpose.

Pittsburgh sits at the confluence of the Allegheny and Monongahela rivers whose waters flow together to form the Ohio River and then into the Mississippi River. We are part of the great Mississippi River drainage that serves nearly one third of the American mainland. This place we call Pittsburgh was a center for commerce, trade, and civilization from ancient times forward. Long before the French and British battled for dominance here at the Point, this was the ancestral land of the Haudenosaunee,

Lenape, Osage, and Shawnee peoples.[34] In preparation for the planned march to the Point, a group of environmental advocates and supporters met at Maren Cooke's Sustainability Salon to meet with some of the elders of the Standing Rock Sioux who had come to perform a water ceremony.

For the indigenous peoples gathered to celebrate the life-force of fresh water, unity with the land, the earth, the sky, and the water is essential for all life. Preserving these elements of Nature is a sacred trust handed from one generation to the next for seven generations. As the water blessing ceremony began, Cheryl Angel spoke most passionately in her own Lakota tongue of the unifying force of water. "Water is life. Without it we cannot live, so it is our sacred duty to protect the water and keep it running free and pure for our time, and for our children and their grandchildren."[35]

The concept of connection to the water, the air, the land is embedded in the civilization of the Standing Rock Sioux, as expressed by Guy Jones, "The drums share the heartbeat of the Earth, our provider. Peace is our goal, at all costs, but we have no peace from the invasion of pollution, from the poison of industries and mining. We have no peace from the taking of the water and the taking of our land."[36] As the ceremony unfolded to the drumming and chanting and dancing of the tribal leaders, water samples brought for sharing were arranged in a circle, co-mingled and blessed, then poured into the rivers' confluence as a symbolic unity with the waters of the entire Mississippi system. Those who stand in solidarity with this ancient ritual are moved with the solemnity and the significance of this tradition—holding sacred the priceless gifts of the living Earth: fresh water, clean air, fertile ground, and the many species that constitute the interconnected web of life.

For the Standing Rock Sioux, the Seminole, the Algonquin, the tribes of the Iroquois Nation, humans are essentially part of nature, not dominant over nature. The sufficiency of all in the community depends on the interdependence of each person. Each contributes for the benefit of the whole, and the community is celebrated as a unit. Decisions honor the ways of the past, recorded in the wisdom of Elders, and consider the implications for seven generations forward. The obligation to protect the water, the land, the resources of Earth is a sacred duty. They speak for all the people who rely on fresh water as a critical need for life.

As we were singing and participating in this solemn ceremony, an armed Coast Guard gunboat began cruising around the Point, back and forth with guns aimed at the crowd, and armed men with binoculars watching. The elders asked the white brothers and sisters to form a circle of protection facing the river between the gun boats and the ceremonial celebrants. We made a double lined cordon along the edge of the rivers holding signs that said "Water is Life" and we kept vigil. This time, the peaceful ceremony was completed without violence, but those protecting the Standing Rock Sioux lands from the incursion of an oil pipeline into their watershed have not been so fortunate.

By contrast, the shale gas industry sees water as a component of production, taken for free from the surface waters at a tremendous rate. Each event of hydraulic fracturing in a deep shale well takes 500,000 gallons of fresh water. The coexistence of abundant water resources with the deep shale seams of the Marcellus and Utica gas deposits makes the Western Pennsylvania corridor attractive for this industry.

As the increase in frequency and severity of storms in the Gulf Coast has damaged or destroyed infrastructure for petrochemical production, the industry scans north and east to this region for the resources it needs to produce gas and plastic. The federal and state rules that strive to protect water—The Clean Water Act, Safe Drinking

Water Act, Resource Conservation and Recovery Act and several others—were suspended for hydraulic fracturing industries by the "Haliburton Loophole" in the National Energy Act of 2005. This permits the process of slick-water hydraulic fracturing to inject water laced heavily with salt (1,300 times more concentrated salt than sea water) and a cocktail of chemicals and fine sand that make it possible to extract gas from deep underground. Water that comes up to the surface with the gas becomes heavily contaminated not only with the chemicals introduced but also with materials extracted from the deep shale formations, including radioactive boron, hydrocarbons, and minerals.

To the petrochemical industry, the water has value as a cheap production element to be used and discarded with minimum concern for the by-products. The methane (natural gas) produced from fracking is used for heating and cooking, and large amounts are liquefied and set for export to other countries. The liquids (ethane and other components of "wet gas") are destined for the petrochemical industry, a much more lucrative undertaking that makes polypropylene plastic pellets, the precursors to many products such as food packaging, film, trash bags, diapers, toys, crates, drums, bottles, food containers, and housewares.

Fracking and the petrochemical buildup to produce plastics threaten prospects for addressing global warming and global pollution. The plastics industry based on raw materials extracted from fossil gas deposits will accelerate both global warming by producing tons of greenhouse gas emissions throughout the production cycle, and the product of this operation will contribute tons of plastic materials to the waste stream contaminating the oceans and landfills. Industry analysts and labor unions look at the immediate jobs scenario, projected to last for 20 years with escalating levels of manufacturing associated with the petrochemical operations. The cost of this industrial expansion is expressed in terms of the climate impact, environmental degradation, and worker and public health and safety deterioration.

The regulatory policies applied to fracking for natural gas and petrochemical production support destruction of the natural environment. Environmental damage to land and ecosystems is the inevitable consequence of the fracking process from the hydraulic fracturing itself, the pipelines, separation facilities, and transportation and production infrastructure. Fracking fragments forests, compromises wetlands, destroys watersheds, emits methane and fugitive hydrocarbon pollutants, and contaminates land with deposits of particulates, radioactive material, and organic compounds.

Removing millions of gallons of fresh water from the surface flows of rivers and streams of Pennsylvania to pump underground for hydraulic fracturing will have long-lasting consequences for the geology of the area. Stream paths will be redirected, groundwater recharge rates will be affected, and plumes of heavy salt will travel through the hydrology of the area. Unknown consequences of this massive redistribution of the water flows will impose both tangible costs to communities as they struggle to assure safe drinking water supplies and indirect costs in the loss of functioning ecosystems. Taxpayers and communities will pay for the transient profits of these multi-national corporations for generations into the future.

The single Shell Appalachia petrochemical plant under construction in Beaver County will send out enough pollution and greenhouse gasses to totally obliterate the climate action efforts of the entire surrounding area. The complex of such facilities touted at the Shale Insight Conference would doom this whole area to a future devoid of hope for reaching any meaningful response to the climate crisis and the global plastics pollution crisis that is compromising the very existence of life on earth.

The profits will enrich a few multi-national corporations, who will pay as little as possible to their workers, and as little as they can manage to the communities, with extended tax credits and subsidies from the taxpayers of the state and nation. Meanwhile, the costs of failing to address or mitigate the effects of climate change extract an enormous cost from all of us. The costs of continuing this intense investment in extending the fossil-based industries far into the future will have catastrophic effects on our ability to mitigate climate change in our communities.

This area has seen the boom-and-bust cycle repeatedly over its history, most recently in the dramatic decline of steel and heavy manufacturing in the 1970 to 1980 decade. The population fell, unemployment reached 25% among those who stayed, and the city was close to bankruptcy. This was an industrial transition without a plan, without consideration of the social and environmental justice issues, and without compassion for the human suffering. The trauma and scars of that time run deep and linger to this day. People see system change as fraught with danger. The myth that air and water pollution are inevitable, if not necessary, side effects to having good jobs is well entrenched in the culture of Pittsburgh.

We can take many important lessons from the trauma of that sad time: First is to recognize that powerful industries will shape the social conditions for their success without regard for the impact on individuals, communities, or future citizens. They will shape the laws to their advantage for as long as they can. Taxpayer subsidies originally applied to encourage the public convenience and necessity of certain enterprises have not been reviewed and evaluated for current conditions. Do subsidies to multi-national corporations with annual profits in excess of the gross domestic product of many countries really support a public convenience and necessity when the result is increased pollution and global warming?

Second, workers are rarely paid what they deserve;

they are paid what they negotiate. Unions had a major role in obtaining fair wages, safe working conditions, and humane hours through battle with the barons of the industrial revolution. Now, the strength of unions has been undermined and eroded by the same forces that offer good wages to workers enduring the effects of lax environmental and public health and safety regulations. The fraught labor movement has alienated entrepreneurs and innovative companies emerging in the renewable energy arena, and in the high technology industries as well. Do we need a better model for determining a fair wage, or is there another way to reflect the value for "the public convenience and necessity" in moving away from a fossil-based economy?

Third, we must recognize that *the laws of nature are not negotiable.* The laws of chemistry, physics, and the biological responses to changes in the physical environment cannot be changed by human declarations or wishes. Our laws and actions must conform to the laws of nature, or we will join the growing list of living things that are going extinct in the face of a warming planet. Increasing greenhouse gas concentrations in the atmosphere raise the temperature of the planet by holding the sun's radiation close to the surface. Increasing concentrations of carbon dioxide in the atmosphere cause increased acidity in the oceans as the carbon dioxide absorbs into the water. Changes in the currents of air and water affect the weather patterns and the water cycle in ways that disrupt established patterns of land use and human habitation as well as habitat for creatures all over the globe.

We face a critical inflection point where the ecological balance shifts to a new equilibrium that is hostile to life as it has evolved over the last five million years. The invisible hand of the market will not drive the fundamental transformation that is required to maintain climate conditions in a temperature range that supports today's living things, including humans. We must apply the moral judgment to make decisions that will sustain a living planet for the future. Where we choose to make investments and what we choose to enable by law will determine the fate of our children.

The greatest tragedy of the Haliburton Loophole is the huge diversion of capital and expectations from a path that sustains life to one of destruction. The enormous subsidies and incentives showered on the multi-national corporations of the petrochemical industry foreclose investment in sustainable, resilient development initiatives within communities. This lost opportunity cost of the fracking/petrochemical industry is a moral decision to destroy the future rather than to preserve it.

When empowered to decide on what kind of future is desired for communities, people develop exciting plans. We face two visions for the future—one where preserving fresh water symbolizes a civilization that recognizes the value of the living Earth and preserves it as the provider of our life support system and one where water is a production medium and land is to be exploited for transient profits. I close with Rachel Carson's prescient comment at the end of *Silent Spring*:

> We stand now where two roads diverge. But unlike the roads in Robert Frost's familiar poem, they are not equally fair. The road we have long been traveling is deceptively easy, a smooth superhighway on which we travel at great speed, but at its end lies disaster. The other fork of the road – the one less traveled by – offers our last, our only chance to reach a destination that assures the preservation of our earth.[37]

The pathways to a sustainable future and the technologies we need to pursue them are at hand. We face an existential decision. Will we leave a legacy of a living earth for our children, or will we remain focused on immediate profits and condemn our children to a future hurtling toward certain destruction of life? Imagine what we can accomplish if we invest in the future instead of subsidizing the past.

4.3 LESSONS FROM
THE HIBAKUSHA

*Presentation to remembering Hiroshima Imagining
Peace on February 4, 2020*

Seventy-five years ago, the United States dropped two nuclear bombs on Hiroshima on August 6, 1945 and Nagasaki on August 9, 1945. Both cities were reduced to rubble, and a shock wave blast area and fire spread over 2.2 miles, with the lethal area extending to a 1.3 miles radius from the point of contact. The justification for this act rested on ending Japan's involvement in World War II and bringing a rapid conclusion to the fighting. Debate over whether this was justified and necessary continue among strategists to this day, but the human suffering and legacy of destruction lingers to this day as a warning against ever deploying nuclear weapons again. The survivors of this bombing, known as the Hibakusha, leave four lessons for our time.

1. The resilience of the human spirit

Imagine waking to the horror of a post–atomic bomb site. The prospect is daunting—infrastructure gone, communication gone, relatives left without knowing the fate of loved ones. Death estimates range from 90,000 to 120,000 for Hiroshima and from 60,000 to 70,000 for Nagasaki because exact tolls were not possible. Bodies were vaporized in the blast zone and bodies were washed out to sea in the tides. Many died of radiation exposure within days or months, many hundreds of thousands survived with lingering illnesses such as anemia, ulcers, asthma, brain tumors, thyroid tumors, and leukemia. Yet, 120,000 volunteers participated in the Life Span Study of Radiation conducted by Radiation Effects Research Foundation, jointly funded by the U.S. and Japan. Most of what is known today about the long-term health effects of radiation has come out of research with those survivors. Dennis Normile reports in *Science*,

> Within six weeks of the bombings, three U.S. and two Japanese expert teams were at work in both cities to study the biological impact of the radiation. Their objectives differed. The Japanese were primarily trying to understand the medical effects on survivors. The Americans wanted to know how and why people died from atomic blast radiation. That might help triage victims—separating those who might be saved from those doomed to die—during future nuclear wars."[38]

Much of the suffering persists long after the initial acute event. The fear of residual genetic effects passed to future generations remains a concern of many Japanese. The discrimination against the hibakusha (survivors of the bomb blasts) persists from the fear that children will be genetically impaired. Research and studies of children born to mothers who survived the bomb have not reassured the public. So, the emotional harm continues long after the event.

But some things cannot be destroyed. As a people, the Japanese show resilience, keeping the memory of the atomic bomb as a herald for peace. Love and hope can thrive in community, even as we struggle together for a better future. The devastation of Hiroshima and Nagasaki stands as a permanent testament to the destructive power of human ingenuity turned to making war instead of peace. The remembrance of this terrible event serves as a spur to peaceful resolution of conflicts.

2. The ethical choice to use nuclear science for benefit rather than for harm

Marie and Pierre Curie discovered polonium and radium, and Marie championed the development of x-rays after Pierre's death. Curie won two Nobel Prizes, for physics in 1903 and for chemistry in 1911. She was the first woman to win a Nobel Prize as well as the first person—man or woman—to win the prestigious award twice. She remains the only person to be honored for accomplishments in two separate sciences.

During the First World War, Marie Curie saw many soldiers die or lose limbs from injuries that were not life threatening but could not be accurately diagnosed in battle conditions. She put together mobile x-ray machines that could be taken to medical centers in the battlefield to allow broken bones to be set, and accurately locate shrapnel and bullets for surgical removal. It was her dream to see x-rays bring many improvements to the practice of medicine. Indeed, the legacy of nuclear medicine has taken this path. Modern diagnostics have advanced to a high degree of sophistication, with surgical procedures simplified through nuclear imaging. Using focused radiation beams to shrink tumors and treat surgically inaccessible lesions has advanced cancer treatments in many areas.

The choice to turn nuclear technology to the destructive force of a bomb was touted as a great scientific achievement. In speaking of the Manhattan Project that produced the atomic bomb, President harry Truman said, "What has been done is the greatest achievement of organized science in history. It was done under high pressure and without failure."[39] Using nuclear science to develop an atomic bomb turned the world on a path of nuclear arms development and containment that preoccupies the global balance of power to this day.

3. The legacy of high-level nuclear waste

High-level nuclear waste is a concern because these materials remain radioactive and can cause health harms to living things. The biological effects of plutonium and other man-made alpha-emitting transuranic elements are primarily dependent upon their entering the body and being deposited in radiosensitive tissues, especially through inhalation.[40] These high-level radioactive materials decay over very long time periods, thus remaining radioactive for thousands of years. For plutonium[239], the half-life is 24,400 years, which means that after that time, half of the radioactivity will remain; for plutonium[242] the half-life is 379,000 years.[41] These high-level radioactive materials are created in weapons production, deployment, testing, and in nuclear power reactors. They are thus man-made elements not found in nature.

At the end of World War II, the Cold War advanced an escalating battle of deterrence that has defined the nuclear age. In the 1950s and into the 1990s, open-air testing of nuclear weapons was established at the Nevada Test Site (NTS). Nuclear weapons testing at the Yucca Flats (NTS) began with a one-kiloton-of-TNT (4.2 TJ) bomb dropped on Frenchman Flat on January 27, 1951. Over the subsequent four decades, over one thousand nuclear explosions were detonated at the NTS.[42] Underground nuclear testing (951 explosions) continued due to public health concerns about radioactive fallout. The westerly winds carried the radioactive plume over Utah where elevated occurrences of cancers were observed. Elevated levels of leukemia, lymphoma, thyroid cancer, breast cancer, melanoma, bone cancer, brain tumors, and gastrointestinal tract cancers were reported from the mid-1950s through 1980.[43] The build-up of nuclear arms has created an eternal legacy of high-level nuclear waste managed at the Hanford Nuclear Reservation.

The Hanford Nuclear Reservation was the site of the Manhattan Project atomic bomb production. The Hanford site was home to the first full-scale production reactor to produce weapons-grade plutonium used in the atomic bomb. During the Cold War, the project expanded to include nine nuclear reactors and five large plutonium processing complexes, which produced plutonium for most of the more than 60,000 weapons built for the U.S. nuclear arsenal.[44]

Nuclear technology developed rapidly during this period, and Hanford scientists produced major technological achievements. Many early safety procedures and waste disposal practices were inadequate, and government documents have confirmed that Hanford's operations released significant amounts of radioactive materials into the air and the Columbia River. The weapons production reactors were decommissioned at the end of the Cold War, and decades of manufacturing left behind 53 million U.S. gallons (200,000 m³) of high-level nuclear waste.[45] In 1989, the Hanford site was declared a superfund toxic site and is under management for cleaning up the 56 million gallons of high-level nuclear waste now in repository there. Radiation leaks from this facility have occurred frequently and numerous lawsuits are in progress surrounding the operation of this high-level nuclear waste facility.

A second initiative of the Cold War was the development of "Atoms for Peace." Launched by President Eisenhower, this initiative had two aspects, one successful and one abandoned almost immediately. President Eisenhower characterized the atoms for peace initiative:

> To the making of these fateful decisions, the United States pledges before you—and therefore before the world its determination to help solve the fearful atomic dilemma—to devote its entire heart and mind to find the way by which the miraculous inventiveness of man shall not be dedicated to his death but consecrated to his life.[46]

Operation Plowshares from 1962-1965 was a series of nuclear tests at Yucca Flats in Nevada. Proposed applications for controlled nuclear explosions included the creation of harbors, canals, open pit mines, railroad and highway cuts through mountainous terrain, and the construction of dams. The radioactive fallout from such uses would have been extensive. Public concerns about the health effects and a lack of political support eventually led to abandonment of the concept.

Nuclear power "tamed" the atom to produce electricity in nuclear fission reactors. In promoting this technology, Lewis L. Strauss, chairman of the Atomic Energy Commission testified to Congress in 1954 that "nuclear power will make electricity too cheap to meter."[47] But despite all assurances and encouragement, industry was skeptical and apprehensive. Finally, Congress passed the Price Anderson Act of 1957 which limited required operator insurance and capped liability in case of accidents. The value of this ongoing federal subsidy to the nuclear industry exceeds $100 billion dollars. Nuclear power plants have supplied about 20% of total annual U.S. electricity since 1990. The 97 operating nuclear reactors in the U.S. produce more than 2,000 metric tons of radioactive waste a year, according to the Department of Energy—and most of it ends up sitting on-site because there is nowhere else to put it.[48]

This legacy of high-level radioactive waste from man-made materials is the burden this nuclear age, opened with the bombing of Hiroshima and Nagasaki, is imposing on our children for millions of years into the future. The development of nuclear weapons and nuclear power without addressing the moral obligation to safely manage and contain the waste is a failure of responsibility for our actions on a grand scale.

4. Nuclear Medicine

The use of nuclear materials in medicine shows the balance between the potential for harm and the potential for benefit. The x-ray has become a standard diagnostic tool for broken bones, dental evaluation, guiding surgical procedures, and evaluating lung diseases. Diagnostic nuclear medicine involves the use of radioactive tracers to image and/or measure the global or regional function of an organ. And, the focused use of radiation has been used for the treatment of tumors to reduce them for better surgical outcomes or to control their growth in areas which are not amenable to surgery. Nuclear medicine is now a $1.7 billion industry. The Society of Nuclear Medicine estimates that 20 million nuclear medicine procedures are performed annually in the United States, of which 12 million are procedures approved for and reimbursed by the Center for Medicare and Medicaid Services.[49] Nuclear medicine has advanced on many fronts, and in this field, the vision of Marie Curie for beneficial uses of radiation sees fulfillment.

Hear and honor the Hibakusha

The Hibakusha have shown the true grace of an oppressed people. Their dedication to contributing to the understanding of radiation effects on health has continued now into second and third generations of studies. Their call for a constant remembrance of the horrors unleashed by nuclear weapons cannot be ignored or forgotten. It is the moral responsibility of all people of our generation to secure the future for all of the children of the 21st century. We must strive for peace.

Treaties and agreements to limit nuclear war emerged soon after World War II. Negotiated between 1965 and 1968 among eighteen nations and sponsored by the United Nations, the initial nuclear Non-Proliferation Treaty was fully executed in 1970 and held for 25 years. It was extended in 1995, with all participants committed to extend the treaty indefinitely. The International Atomic Energy Administration was established to enforce compliance. As of August 2016, 191 nations have signed the agreement, including the U.S. North Korea withdrew; India, Israel, and Pakistan did not sign and all have nuclear weapon capability.

The Comprehensive Nuclear Test Ban Treaty Organization, organized under the sponsorship of the United Nations, notes that 184 countries have ratified the Nuclear Test Ban Treaty. Eight more will put it in permanent effect to ban nuclear weapons testing forever. "We must remain committed to the Comprehensive Nuclear-Test-Ban Treaty's entry into force," says CTBTO head, Lassina Zerbo.[50] At this point, there are destabilizing elements at play in nuclear arms threats in several countries around the world involving the United States, Russia, Iran, and North Korea. This is a complex area of international power jousting, one that must remain confined to the verbal stage for the sake of our survival as a species, and as civilizations.[51]

We can each play a part in securing the future. We must insist on funding and attention to managing the existing high-level nuclear waste repositories. We must recognize that nuclear energy use includes an obligation for thousands of years for waste management—now in temporary storage at 97 reactor sites all around the country. We must demand accountability from our leaders to strive for peace rather than to escalate nuclear weapons capabilities. We can learn from the Hikabusha that we are human—resilient, enduring, and capable of great empathy.

Pray for Peace
Work for Justice
Dance for Joy

4.4 RE-IMAGINE AMERICA IN HARMONY WITH NATURE

Earth Day presentation April 22, 2020

We are in a state of emergency, not only because of the COVID-19 pandemic but also the ongoing and escalating existential crises of global warming and global pollution, especially from plastics. Solving this trio of global crises will require collaboration, community, and a sense of commitment to the future. Our country is deeply divided and out of balance in response to any single crisis, totally rudderless and struggling to address these overlapping issues.

But sometimes, addressing a constellation of crises together brings solutions closer. This is especially true when the underlying causes overlap, and so do the solutions. The story of modern civilization since the Industrial Revolution has rested on subjugating nature through resource extraction, commercial agriculture exploiting the land, and piecemeal implementation of mitigation strategies. This moment in time offers an opportunity to reset our trajectory. We can reimagine America in a path that flows in harmony with Nature.

Rachel Carson wrote the following words in 1946, but her admonition rings truer today than ever:

> Because it is more comfortable to believe in pleasant things, most of us continue today to believe that in our country there will always be plenty. This is the comfortable dream of the average American. But it is a fallacious dream. It is a dangerous dream. Only so long as we are vigilant to cherish and safeguard our resources against waste, against over exploitation, and against destruction will our county continue strong and free.[52]

The emphasis on economic outcomes above all else as both a metric for progress and a guide for public policy has torn great holes in the social safety net and shredded basic environmental protections for clean air, fresh water, fertile ground, and the biodiversity of species. These gifts of the living Earth, the ecosystem services, do not count in the Gross Domestic Product. These are our life support system that policies based heavily on economic profits to corporations are destroying. We need to change direction and restore the balance among economy, environment, and society. Three important measures of the well-being of the nation illustrate the urgent need to change direction: the health of the people, the condition of the environment, and the distribution of wealth.

State of American health

The COVID-19 pandemic has placed a sharp spotlight on the failings of our health care system. According to Cynthia Cox, Director of the Peterson-Kaiser Health System,

> The U.S. performs worse than average among similarly large and wealthy countries across nearly all measures of preparedness for a pandemic.... The coronavirus outbreak is already exposing inefficiencies and inequities in our health system. Compared to most similarly large and wealthy countries, the U.S. has fewer practicing physicians per capita but has a similar number of licensed nurses per capita. Looking specifically at the hospital setting, the U.S. has more hospital-based employees per capita than most other comparable countries, but nearly half of these hospital workers are non-clinical staff. Overall, the U.S. has fewer hospitals and hospital beds per capita compared to other similar countries.[53]

In a comprehensive study of U.S. health

parameters compared to other developed countries, Americans showed inferior health standards in nine categories including adverse birth outcomes, obesity and diabetes, chronic lung diseases, heart disease, and injuries and homicides.[54] The causes for these poor health outcomes are varied. For example, the U.S. health system is highly fragmented, with limited public health and primary care resources and a large uninsured population. In 2018, 8.5 percent of people, or 27.5 million, did not have health insurance at any point during the year, an increase of 7.9% over 2017 levels.[55] Compared with people in other countries, Americans are more likely to find care inaccessible or unaffordable and to report lapses in the quality and safety of care outside of hospitals.

The U.S. has an infant mortality rate of 5.8 per 1,000 live births whereas the average comparable developed country has an infant mortality rate of just 3.4 per 1,000 live births.[56] The U.S. maternal mortality rate was 29.9 per 100,000 live births in contrast with just 6.1 per 100,000 in the average comparable developed country. According to the CDC, black women are three to four times more likely to die from a pregnancy-related complication than non-Hispanic white women.[57] The overall U.S. rate of pregnancy-related mortality has been rising over two decades, while in the rest of the world it has been going down.

According to the National Research Council, many conditions that might explain the U.S. health disadvantage—from individual behaviors to systems of care—are also influenced by the physical and social environment in U.S. communities.[58] For example, built environments that are designed for automobiles rather than pedestrians discourage physical activity. Patterns of food consumption are also shaped by environmental factors, such as actions by the agricultural and food industries, grocery store and restaurant offerings, and marketing.

U.S. adolescents may use fewer contraceptives

because they are less available than in other countries. Similarly, more Americans may die from violence because firearms, which are highly lethal, are more available in the United States than in peer countries. A stressful environment may promote substance abuse, physical illness, criminal behavior, and family violence. Asthma rates may be higher because of unhealthy housing and polluted air. In the absence of other transportation options, greater reliance on automobiles in the United States may be causing higher traffic fatalities. And when motorists do take to the road, injuries and fatalities may be more common if drunk driving, speeding, and seatbelt laws are less rigorously enforced, or if roads and vehicles are more poorly designed and maintained. Tragically, the U.S. ranks last in overall health care among developed nations.[59]

State of the U.S. Environment

This decade has seen the advance of global warming from the ends of the earth to the closest neighborhood. But even as the scientists who monitor climate change and document the destruction of oceans, wetlands, forests and grasslands escalate their alarms, the Trump Administration relentlessly attacks the policies and protections that could help to slow the pace and mitigate the worst of the harms. The U.S. is scheduled to withdraw formally from the Paris Climate Accord at the end of 2020 and the current U.S. emissions trajectory is totally inadequate to be on track for preserving the target of controlling emissions to no more than a two-degree Celsius rise above 2005 levels.[60]

With 48 of his targeted 95 environmental protections rescinded, reversed, or weakened as "burdensome" curtailments of business, the deliberate destruction of vast parts of the life support system of this country seems assured. Safe drinking water from rivers, streams, and wetlands once protected by law now face unrestricted pollution from reckless corporate interests. With regulatory and administrative agencies headed by lobbyists and executives from coal, oil, natural gas, mining, logging, and commercial agricultural interests, there is no advocate in the government to protect the public interest or to safeguard the future.[61]

Now in the face of the COVID-19 pandemic, the EPA is stopping enforcement of environmental regulations and accelerating permits for oil, gas, and coal projects. The wealth of the land is being stripped away and imperiled forever at the behest of special corporate interests whose next quarter profits drive relentlessly to destruction. After decades of decline in air pollution nationwide, increases of 5.5% in small particulate pollution has occurred since 2018.[62] The uptick in emissions comes from rollback in environmental protections, increased wildfires, and a rise in driving and natural gas use, obliterating the curtailed emissions from declining coal.[63]

State of American Prosperity

The income equity gap in the U.S. is wider than ever and has been accelerating. While Americans enjoyed a shared growth in prosperity from the end of World War II into the 1970s, the gap between the most wealthy and least wealthy has been growing, with fewer and fewer Americans at the high wealth end of the scale. Since 1970, average income after transfers and taxes quadrupled for the top one percent of the distribution; the increases were much smaller for the middle 60 percent and bottom 20 percent of the distribution.[64]

The federal government spends more than $400 billion to support asset development, but those subsidies primarily benefited higher-income families and racial wealth disparities continue. About two-thirds of homeownership tax subsidies and retirement subsidies go to the top 20 percent of taxpayers, as measured by income. The bottom 20 percent, meanwhile, receive less than one percent of these subsidies. Blacks and Hispanics, who have lower average incomes, receive much less of these subsidies than whites, both in total amount and as a share of their incomes.[65]

Low-income families benefit from safety net programs, such as food and cash assistance, but most of these programs focus on income—keeping families afloat today—and do not encourage wealth-building and economic mobility in the long run. What's more, many programs discourage saving: for instance, when families won't qualify for benefits if they have a few thousand dollars in assets or when they have to give up rent subsidies to own a home; and tax subsidies highly favor investments over savings.[66]

We Can Be Better Than This!

> Unaware of what we have done or its order of magnitude, we have thought our achievements to be of enormous benefit for the human process, but we now find that by disturbing the biosystems of the planet at the most basic level of their functioning we have endangered all that makes the planet Earth a suitable place for the integral development of human life itself. Thomas Berry[67]

The COVID-19 pandemic experience gives us an opportunity to re-think what we are doing. We have a chance as we are forced to pause in the headlong pursuit of daily endeavors to sharpen focus on what truly matters. The stark reality of the inequities in our economy, our social network, and our environment cannot stand if we are to build a more resilient, a more equitable and just, and a more environmentally healthy society.

The Re-Imagine movement has been working through Pennsylvania communities, facilitated by the League of Women Voters, bringing ideas from the ground up for a new way forward. We can take lessons from the Re-Imagine movement to date and build a vision for the future that will bring a more equitable, healthy, and prosperous time for our country. Many of the sessions in Pennsylvania begin with the thought exercise: "Imagine what you would do for economic development in your community if you had the $1.6 billion in incentives given to the Shell Appalachia petrochemical plant?"

The answers to this question from Beaver County to Erie, Lehigh Valley, Johnstown, Wheeling, West Virginia and the Re-Imagine Appalachia effort covering Pennsylvania, Ohio, West Virginia, and Kentucky reveal enormous creativity and aspiration for a better way forward. On this 50th anniversary of Earth Day, I am moved to imagine what America could be like if we break the bond of our economy to the extractive fossil industries that are killing our planet and making us sick. As we plan for life after COVID-19, we can reset our way of life to a new, better way forward that addresses the trio of crises all at once.

Principles and Values

This period of a global pandemic requires isolation and separation from public gatherings. Parents keep distance from their children and grandchildren; neighbors wave instead of sitting together; merchants desperately seek ways to keep connection with clients and customers. Many fall into despair as isolation shrinks our world. The need for distancing sharpens our sense of interconnectedness and interdependence. It is important to reflect on priorities and assess what really matters in our life.

I share here the vision of Donella Meadows who wrote in *The Limits to Growth* a scenario where we avoid collapse of our civilization:

> People don't need enormous cars; they need respect. They don't need a closet full of clothes; they need to feel attractive and they need excitement, variety and beauty. People need identity, community, challenge, acknowledgment, love, joy. To try to fill these needs with material things is to set up an unquenchable appetite for false solutions to real

and never-satisfied problems. ... A society that can admit and articulate its nonmaterial needs and find nonmaterial ways to satisfy them would require much lower material and energy resources and would provide much higher levels of human fulfillment.[68]

I have great confidence in the basic common sense of people when they have a chance to choose a better way forward. We can advance a better path, one that redirects the destructive impetus of the past and opens a new epoch of human endeavor where we align our economy and our society to exist in harmony with the natural systems of the living Earth that are our life support system. If we shift from exploitation to nurture as the operating premise of our relationship to the earth's ecosystems, we will build a better future, one that offers hope and joy to our children instead of escalating despair.

Recognize the Universal Rights of Mother Earth

The earth is a living system of which humans are but one part, not property to be owned or destroyed for human profit. The laws of nature co-evolved over millions of years—chemistry and physics, and biological and physiological responses to conditions in the environment—define complex inter-relationships among all living things and connect the living earth elements with the mineral and inert elements. These functions are inherent in living systems, priceless attributes of the living earth that are not reflected in the drivers of the economy. To achieve meaningful and lasting solutions to the existential crises of global warming and global pollution, the laws of nature must be incorporated into the practices of civil society. The laws of Nature are *not* negotiable!

Indigenous peoples all around the world have long recognized the necessity of living within the laws of nature and do so by respecting the rights of the living earth. A gathering of indigenous peoples in the People's Climate Conference at Cochabamba, Bolivia on Earth Day 2010 adopted a Universal Declaration of the Rights of Mother Earth that was introduced at the COP-15 meeting in Paris in 2015, arguing for a 1.5°C increase above pre-industrial ceiling for global warming. The goal of a 1.5°C ceiling was incorporated into the Paris Climate Accord of 2015 and was signed by 195 nations due to these efforts. The justification for this action states in part:

> We the Peoples and Nations of Earth are all part of Mother Earth, an indivisible, living community of interrelated and interdependent beings with a common destiny; and… Recognizing that capitalism and all forms of depredation, exploitation, abuse and contamination have caused great destruction, degradation and disruption of Mother Earth, **putting life as we know it today at risk** through phenomena such as climate change establish this Universal Declaration of the Rights of Mother Earth.[69]

Re-Imagine America in Harmony with Nature

There are no technological barriers to making rapid and meaningful changes toward sustainable climate solutions. All the technologies necessary to address the major sources of global warming and global pollution are in hand and will only improve in effectiveness as they become more widely adopted. Only the political will to act stands in the way of transforming our economy. More than 73% of Americans want action on climate change but are deeply divided on partisan lines (67% of Democrats and 21% of Republicans).[70] The U. S. Constitution vests the power of government in the people. We **have** the power to act to save our world.

We stand at a crossroad now. In one direction, we

can continue toward a future based on petrochemical industries—build out the infrastructure that will bind our economy to natural gas and plastics for another fifty years. Or we can recognize the ultimate futility of this pursuit and turn our investments, our education tools, our might and political will toward building a sustainable future. The tools for doing this are at hand:

- renewable energy systems;
- regenerative agriculture that captures carbon and restores the fertility of the land;
- non-fossil-based materials in a circular supply chain; and
- preserving the biodiversity of the earth in living ecosystems that provide fresh water, clean air and fertile ground.[71]

When communities work together to examine how to reshape future growth, people focus on renewable energy systems, whether it be to reclaim blown off mountains or to enhance the productivity of farmlands with solar arrays and wind systems. People want to grow hemp, flax, and bamboo with local manufacturing to convert plant materials to goods. Organic farming in both urban and traditional agricultural areas gains popularity. Local manufacturing and re-manufacturing surfaces frequently as an important way to renew communities.

All these initiatives focus also on preserving and enhancing the features that contribute unique character to communities, that preserve the specialness of place and identity.[72] If the government infrastructure can be aligned to support and empower community plans, the innovation and resilience that can emerge will become a platform for a new America, an America where communities come together as part of the land.

We Are Facing an Emergency: Climate, Pollution, and Pandemic

On this 50th Anniversary of Earth Day, we must recognize the true existential crises we face from human activities that destroy the natural systems of the living earth. We must make a u-turn in our policies. This requires a level of commitment equivalent to the mobilization of World War II. The tools are at hand. For 2020, these priorities can drive progress:

1. **Stop subsidizing fossil fuels** research, exploration, production, processing, and use. Taxpayer dollars in the U.S. alone exceed $649 billion annually in direct subsidies. Replace this with a bottom-line tax deduction for all property owners for energy efficiency, renewable energy installations, carbon sequestration in trees and organic farming, and replacements of fossil resources with non-fossil materials such as bamboo, hemp and algae.

2. **Reverse the primacy of mineral rights over surface rights.** Ecosystem services such as wetlands, grasslands, and forests depend on intact surface conditions. Disruptions for mining, drilling, excavation, and erosion destroy the ecosystems that provide our life support.

3. **Re-invest in communities.** Give communities the resources to plan for a diverse and stable future based on renewable resources while affirming community values. Invest in people, rather than multi-national corporations with no allegiance to sustainability.

4. **Protect and care for the people who are victims of social and humanitarian disruptions associated with the response to climate change**. For the

workers of the oil, gas, and coal industries, transition to productive jobs in the new economy, protecting pensions and health benefits, and maintaining the dignity of their worth are essential. Millions of people are thrust into forced migration from climate effects around the world, and even within the U.S. Criminalizing people who face extended drought and social collapse is inhumane and demeans our humanity.

The transformation to a society living in harmony with nature will place priority on protecting biodiversity in all areas of the world as an excellent indicator for the health of the complex ecosystems that comprise Earth's life support system. We are facing a critical time in which we will choose the fate of our living Earth for hundreds of years into the future. In making the critical choices about energy and all resource management, we must place greater value on the living things, rather than the dominant drive of profits to corporations. The plan for a just transition must address the needs of people caught in the transition—the oil, gas, and coal workers especially—who will need to transition to new ways of working in new fields. Just and equitable solutions will need to include protecting pensions and health benefits while re-training for workers in transition.

A companion to a policy of reinvesting in communities can empower people to restructure our society with a more diverse and locally responsible economy. We need a new system of governance that relies on a doctrine of public trust for natural resource management. In such a system, common resources are managed for the long-term benefit of the whole of society collectively, rather than to be owned and divided to profit individuals or corporate owners. A specific major change in this approach would place the ecosystem services, mostly on the surface of the earth, as priority for protection above the rights for extracting mineral deposits from deep underground. Such an approach recognizes the priceless value of the services the living earth delivers to all living things for free, as conditions of mutual support. We can transform our society to align with the natural cycles of resource use, recovery, and reuse, rather than changing raw material to trash as rapidly as possible.

The best way to move forward is to remember that we are more alike in our common humanity than different in political stance, race, gender, religion, or culture. If we protect the rights of the living Earth and connect our own fate to the fate of the natural world, we will find the courage to make the necessary changes. The result will be a better future and the legacy of a renewed sense of wonder in the miracle of creation. Each person can act. Each person matters in the great interconnected web of life.

Make your own pledge to Re-Imagine America in harmony with Nature!

4.5 TRANSFORMATION: FROM CONQUERORS TO STEWARDS

Written February 4, 2022

On this cold winter day, the roads are obscured with softly falling snow. In the stark black and white scene, I thrill to hear the mating call of a brilliant red cardinal sitting in the holly bush outside my kitchen door. He sings with confidence in the spring to come and a long summer thriving in the privet hedge where he and his mate built four nests last season, all successful in fledging new cardinals into this backyard wilderness. The cycle of life forecast with joy, beauty, and grace.

I reflect on where we stand in our battle to preserve our life support system in the face of deeply entrenched obstruction. We have lost a strong and steady voice with the passing of Edward Oliver Wilson on December 24, 2021. Ed Wilson sat in my garden and at my dinner table when he was the keynote speaker for the Rachel Carson Legacy Conference on Biodiversity and received the first Rachel Carson Legacy Award in 2008. He was a reflective and gentle man, completely grounded in science as the explanation for human behavior and society's patterns of behavior. His work in sociobiology was controversial and put him at the center of numerous disputes. As he moved into retirement, he spent more time writing, speaking, and advocating for preserving the biodiversity of this marvelous living Earth, our only home. He was realistic about the challenges of climate change and human development but remained optimistic. In a 2012 interview with the *New York Times*, he said: "I'm optimistic, I think we can pass from conquerors to stewards."[73]

The concept of stewardship to preserve half of the earth as wild natural spaces stands as both his legacy and challenge to us. This shift in concept from using the earth as a source of resources to be extracted for economic products to using the resources of the earth in regenerative and restorative ways lies at the heart of the transformation to a sustainable society.

The first transformation necessary for this major shift in approach to the place of humans in the world begins with a change in attitude. We are facing multiple existential crises, all interwoven, all derived from the basic problem of consuming more of the Earth's resources than can be replaced. In addressing this problem, we are not facing a technology problem, but rather an ethical problem—a crisis of moral commitment to preserve the life support system of this planet for our children and for tomorrow's children. We must infuse consideration for preserving and restoring the ecosystem services of the living earth into all our decisions about land use and resource use.

Regenerative Agriculture

Re-configure agricultural lands to restore fertility to the soil, capture carbon, and create resilience against drought and pests. The demand for food is projected to increase by 60 per cent by 2050. The global food system must be transformed to achieve two purposes. First, it must ensure food and nutritional security and put an end to hunger, once and for all.[74] Second, it must reverse the degradation that human actions have caused and restore ecosystems.[75]

Regenerative agriculture is gaining acceptance as large commercial operations suffer from drought, inundation by contaminated flood waters, and crop failures. Regenerative agriculture is a triple-win situation. Consumers can receive healthier foods, farmers can have a more secure and prosperous future, and the planet will benefit because regenerative agriculture provides it a better chance to heal and restore itself.[76] Major food companies are increasing commitments to regenerative practice in their supply chains, with significant increases in the acreage dedicated to regenerative practice for both produce, grains and livestock.

Only five million acres of farmland are dedicated to regenerative agricultural practices in the U.S. today, but major producers have committed large acreage to regenerative practice: WalMart, 50 million acres; PepsiCo, 7 million acres; Cargill, 10 million acres, all by 2030.[77] The trend toward regenerative agriculture holds promise as leaders in Europe and the United Nations have developed broad programs to support farmers in converting from mono-culture industrial operations to regenerative practices.

Consumers Demand Sustainability in Products

Millennials are twice as likely (75% vs. 34%) than baby boomers to say they are definitely or probably changing their habits to reduce their impact on the environment. They're also more willing to pay more for products that contain environmentally friendly or sustainable ingredients (90% vs. 61%), organic/natural ingredients (86% vs. 59%), or products that have social responsibility claims (80% vs. 48%).[78] The post-war (WWII) impetus to buy more stuff and use conspicuous consumption as a mark of status is diminished in the millennial generation.

With greater awareness of the need for sustainability, there is a surging interest in recycling, reuse, and refusing unnecessary products. Globally, 85 percent of people indicate that they have shifted their purchase behavior towards being more sustainable in the past five years. Globally, sustainability is rated as an important purchase criterion for 60 percent of consumers. In the U.S., this number is just over the global average at 61 percent.[79] The challenge in this trend is to make meaningful transparency in sustainability practices, as distinguished from "greenwashing" marketing campaigns. Tech-savvy millennials can access tools easily to check out and verify claims of product sustainability. Consumer pressure is having a significant effect on reducing deforestation and in achieving fair wages for workers at all levels of the supply chain.

Greening Urban Spaces

People resonate with green spaces, whether small parks, treed walkways, or grand forests. Being in or near green spaces contributes to a sense of well-being and promotes mental health. Green spaces help to support ecosystem health in urban spaces, and as urbanization continues across the globe, it is essential to incorporate green spaces into growth areas. There has been a history of disparity in green spaces between wealthier neighborhoods and poorer ones. As the move toward green space incorporated into urban spaces continues, assuring ways to green poorer neighborhoods becomes a central part of restoring environmental justice to disadvantaged communities. Community gardens have contributed a tremendous impetus to green space preservation in urban areas.

According to the Centers for Disease Control and Prevention, food deserts are areas that lack access to affordable fresh fruits, vegetables, whole grains, low-fat milk, and other foods that make up a full range of healthy diets.[80] More than 23 million Americans have to trek more than ten miles to get their nearest supermarket.[81] Urban agriculture—in particular, community gardening—has become increasingly popular in the 21st century, with the number of Americans growing food in community gardens rising by 200% between 2008 and 2016.[82] This initiative has emerged in response to food deserts and also to improving the restoration of older industrial cities, like Detroit, where gardens and open green space are becoming permanent parts of restoring abandoned industrial land.

Small urban gardens often incorporate pollinator-friendly plantings and reduce heat concentration in paved urban areas. As the benefits of green spaces are better recognized, the dispute over land valuation to preserve gardens and forested areas within urban spaces becomes more acute. The tax base for land with buildings exceeds the tax base for land devoted to agriculture so far. Too often a community garden established from the rubble of an abandoned building site is cavalierly dismissed when a developer buys the property or takes it over from title transfer without any regard for the community investment in gardening, sometimes over many years. This is one of the areas where transformative thinking about incorporating the value of green space into property value determinations can improve the sustainability of urban areas.

Design Complete Neighborhoods

With the end of World War II, the advance of motorized travel in the U.S. surged. The apparatus of war shifted quickly to producing consumer goods—automobiles, appliances, big houses, disposable goods from plastic. All these trends contributed to a markedly unsustainable way of living so that now Americans have the highest level of greenhouse gas production per capita in the world. If everyone on earth lived the way the average American does, it would require five and a half planets to support us. We have only one Earth.

Re-constructing the American suburban sprawl mode of living to create more holistic neighborhoods conducive to healthy living with green spaces, walkable connections from work, school, home and recreation seems an impossible pipe dream. However, as the COVID-19 pandemic has introduced lockdowns and working and schooling from home have become almost normal, new priorities emerged. First, the disparity in broadband access reveals a greater economic divide. Nearly 73% of U.S. homes had home-internet access. However, examined by demographic groups, rates of home-internet accessibility show significant disparities based on economic status and race.

Just less than half of households with annual family incomes less than $20,000 have home-internet access.

Furthermore, 83% of Asians and 81% of Whites have home-internet access, compared to 72% of American Indian/Alaska Natives, 70% of Hispanics, 68% of Blacks.[83] Addressing the inequities in access and infrastructure for smooth connectivity to the internet is now a necessity for modern life. Internet access and speed are not only essential to keep up with many aspects of modern society, but they are crucial in bolstering resilience of individuals, families, communities, broader social networks, and overall cohesion. As the federal impetus to increase investment in infrastructure moves forward, it is important to recognize that communities will be more resilient, productive, and sustainable if they have infrastructure that allows more holistic designs of development. Investing in micro-grid centers in business districts, and safe pedestrian amenities like sidewalks, sheltered bus stops, and bicycle lanes can increase the formation of neighborhood resilience.

Re-Wilding the American Dream

The American dream of a two-story, single-family house with a two-car garage on a half-acre of lawn with one tree in the front yard and a white picket fence around it is over. The residential lawn takes up 40 million acres of land, 2% of the land area of the U.S. Americans spend more than $30 billion annually on pesticides, herbicides, and fertilizers and on water to irrigate lawns.[84] Many of the chemicals used on lawns wash into the stormwater system and contribute to endocrine-disrupting chemical pollution of waterways. This vast expanse of lawn could be replaced with native plants, grasses, shrubs and trees to provide wildlife and pollinator refuges throughout the country.

When neighborhoods combine efforts to create wildlife friendly spaces, entire ecosystems can return to formerly barren expanses of green lawn. Even adding vegetable gardens to uninterrupted lawn has benefits. Placing groves of trees at the inner corner of properties (instead of along the street under the power lines) can break up heat islands and provide unfettered growing space for shade trees. Growing permaculture plantings of perennial berry bushes under larger trees can offer a diversity of habitat as well as a diversity of food sources for birds and small creatures. Allowing lawns to grow clover, violets, and other perennial low-growing grasses supports pollinators and, if mowed, can be sufficient for walking paths and garden living spaces. A creative approach to living in harmony with nature emerges from these initiatives.

Whether we live in a high-rise apartment building in the middle of a city or on an individually-owned plot of land, we have opportunities to influence the planning for a greener space. We can make decisions about what we eat, what we use, how we travel, and how we interact with each other, increasing our quotient of treating the world around us as well as the land with respect. Holding the dignity of every person we meet can make space for constructive dialogue. Let us join in the effort to steward our living earth, live lightly on the land, and restore the resilience of the robust ecosystem services that support our life.

We cannot have healthy people in an unhealthy world. We all have a part to play in making our interaction with the living earth a positive experience preserving and regenerating our life support system. The earth, driven by the power of the sun, provides everything we need to thrive—fresh water, clean air, fertile ground, and the wide diversity of species that make up the great web of life, of which humans are but one part. The laws of nature are not negotiable. We must align our laws so our civilization will thrive in harmony with nature, rather than on its destruction.

4.6 A PERSPECTIVE ON RACHEL CARSON'S *SILENT SPRING* AFTER 60 YEARS

This speech was given as the C.F. Reynolds Lecture to the Medical Historical Society of the University of Pittsburgh Medical School on September 20, 2022

Good evening, Ladies and Gentlemen of the University of Pittsburgh Medical Historical Society. Thank you for your kind invitation, and especially, thanks to my colleague and friend Bernie Goldstein, now living in the second home of my heart in Anchorage Alaska.

In October of 1963, but a few months before Rachel Carson succumbed to breast cancer, she gave her last major public address to the Kaiser Foundation Hospitals and Permanente Medical Group in San Francisco. This speech *"On the Pollution of Our Environment"* distilled much of the work of her whole life with a clear admonition of peril ahead if the pattern of dispersing toxic chemicals into the environment were to continue. Carson saw the connection between the health of the environment and the health of people as a manifestation of the great interconnected web of life, of which humans are but one part. This deep, centered conviction was instilled in Rachel Carson from her childhood in Springdale PA, where she grew up with the freedom to explore her boundless curiosity about the world around her. Indeed, this is the first and most essential attribute of a scientist: to know how to observe, and to know how to question and seek answers within a disciplined line of inquiry.

Rachel Carson's science was not that of the experimenter with massive amounts of data subject to rigorous statistical analysis. We are all familiar with the tedium of compiling data and meeting the rigors of statistical validation. Rather, hers was the science of seeing the connections among the intricate parts of living systems and drawing the implications from her observations into the light of public discourse. In this speech, she described herself for the first time as an ecologist, then a relatively new discipline.

> I suppose it is a new and rather humbling thought, and certainly one born of this atomic age, that man could be working against himself. In spite of our rather boastful talk about progress, and our pride in the gadgets of civilization, there is, I think, a growing suspicion—indeed, perhaps, an uneasy certainty—that we have been sometimes a little too ingenious for our own good. In spite of the truly marvelous inventiveness of the human brain, we are beginning to wonder whether our power to change the face of nature should not have been tempered with wisdom for our own good, and with a greater sense of responsibility for the welfare of generations to come.[85]

Rachel Carson brought two central concepts to her work and to her whole life perspective. First this overwhelming sense of precaution in protecting the ways of nature. Perhaps most poignantly noted in her dedication of *Silent Spring* to Albert Schweitzer, quoting his sentiment: "Man has lost the ability to foresee and forestall. He will end by destroying the Earth."[86] And the more hopeful perspective from a personal sense of wonder at the complexity of creation: "Those who contemplate the beauty of the earth find reserves of strength that will endure as long as life lasts."[87]

Rachel Carson's success in persuading even

bureaucratic agencies and elected Congressmen grew from her passion and effective writing. As a child and an early student, she aimed for a career as a writer. Indeed, she was a published author at the age of ten.[88] In college, Pennsylvania College for Women, now Chatham University, she came under the tutelage of Mary Scott Skinker, her biology teacher. In studying biology, Rachel Carson wrote to her mother: "Now I know what I am writing about!"[89] and changed her major to biology. Her credentials as a scientist were questioned when she was giving testimony to Congress after publishing *Silent Spring*. In answer to Senator Gruening of Alaska, she noted her Masters Degree in Biology from Johns Hopkins University, her teaching at the University of Maryland, and years of research at Woods Hole Marine laboratory.[90] This was the site of her only descent under the ocean. As a woman scientist, she focused her attention on the estuarine areas, the edge of the sea, because women were seen as unlucky, even hazardous, on working vessels.

Rachel Carson was not the first to bring science to public policy. Indeed, the nuclear physicists and engineers who drove the success of World War II had greater influence on public policy, in the form of support for atomic energy, and for "atoms for peace" in nuclear power. The hard sciences—chemistry, physics, engineering were never questioned, as innovations ballooned from the war effort and the space quest of the 1960s. What was new with the presence of Rachel Carson in the US Bureau of Fisheries and the Fish and Wildlife Service from 1936 to 1952, was her insistence on looking at the responsibilities of these federal agencies with the wholistic lens of the ecosystem.

During her time working as an aquatic biologist, and later as editor of all agency publications, she wrote strong objections to selective predator removal, urging that wolves and coyotes were essential components of healthy forests and fields. She established the rationale and the program for preserving National Wildlife Refuges.

But for all the people, the preservation of wildlife and of wildlife habitat means also the preservation of the basic resources of the earth, which men, as well as animals, must have in order to live. Wildlife, water, forests, grasslands—all are parts of man's essential environment; the conservation and effective use of one is impossible except as the others also are conserved[91]

Her work was central to establishing laws to protect migratory birds and set the foundation for the Endangered Species Act. As a marine biologist working within the Bureau of Fisheries, she provided a strong voice for habitat preservation and for limitations on over-fishing, set in the framework of the life cycle of the individual species.

Rachel Carson was at the table, in staff meetings, and in meetings evaluating the research of filed biologists.[92] Many issues concerned her in that time, especially the science discovered but mis-applied. She objected to DDT crop spraying with the intention to eradicate the fire ant. She had conversations with Edward O. Wilson about the natural history and life cycle of the fire ant to inform her opinion.[93] She considered DDT to be the equivalent of an agricultural atomic bomb and engaged in the dispute between the Department of Interior which opposed broadscale DDT spraying and the US Department of Agriculture which supported it.

In many ways, the period during and following World War II was a time of great innovation. Intense investment in the munitions of war turned in peacetime to production of plastics, fertilizer, herbicides, and pesticides…the entire petrochemical complex that became the underpinnings of the modern economy. Suburban sprawl not only separated people from farms, with the interstate highway system promoted by President Eisenhower, cars became the medium for the modern age. Farm yields grew as farms consolidated into industrial scale enterprises, and there were fewer farmers on the land. But this movement to consolidating farms into mono-crop industrial operations held disaster for the natural land. Topsoil was destroyed, agricultural biodiversity decreased by 75% in the last 100 years.[94]

Rachel Carson brought her concerns about workers exposure to harmful chemicals to the attention of Labor Secretary Frances Perkins. She was worried about the drift from spraying fields moving into neighborhoods and homes, and she worried about the interactions of long-lasting harmful chemicals with each other and in combination with radioactive fallout from nuclear testing. She was concerned about the unintended consequences of exposing people to a chemical stew, and she was alarmed about harmful chemicals entering the food chain.

Rachel Carson has been accused of being responsible for banning DDT. But that did not take place until after the EPA was established in 1970 and finally acted to ban DDT in 1972. Rachel Carson was primarily concerned about taking caution in using such harsh chemicals without restraint. She wrote in *Silent Spring*:

> It is not my contention that chemical insecticides must never be used. I do contend that we have put poisonous and biologically potent chemicals indiscriminately into the hands of persons largely or wholly ignorant of their potential for harm. We have subjected enormous numbers of people to contact with these poisons, without their consent and often without their knowledge."[95]

This is one example of the way Rachel Carson brought the question of ethical obligations into the discussion of the use of chemicals. In this period of rampant innovation and industrial development, she called out a voice of caution, of concern for unintended consequences. The chemical industry descended upon her with vicious and relentless criticism. From the learned Dr. Robert H. White-Stevens who gave an interview to CBS stating, "If man were to follow the teachings of Miss Carson, we would return to the Dark Ages, and the insects and diseases and vermin would once again inherit the earth."[96]

She endured personal attacks from prominent figures such as William Darby of the Vanderbilt University School of Medicine who thought "If not an outright Communist, surely Carson was linked to 'food faddists' or the organic gardeners, the anti-fluoride leaguers, the worshippers of 'natural foods,' and those who cling to the philosophy of a vital principle, and pseudo-scientists and faddists."[97] Her industry detractors sought to dismiss her concerns as those of a hysterical woman. She was dismissed as a "Nature lover" and was often depicted in bucolic woodland settings with children and wearing binoculars, rather than in a scientific setting in the laboratory with her microscope.[98]

Rachel Carson delved through mountains of materials in researching Silent Spring. Carson studied the work of many scientists, both proponents of pesticides who felt harm had not been sufficiently documented, like Wayland Hayes and those who thought there were damages but that they could be avoided by alternatives to spraying, like George Wallace. There are 52 pages of scientific citations in *Silent Spring*—heavy peer-reviewed journals like *Clinical Toxicology of Commercial Products, Archives of Industrial Health, Proceedings of the Society of Experimental Biology and Medicine, American Medical Association Archives of Industrial Hygiene and Occupational Medicine.*

Scientists were ignored as the use of DDT spread. Some of the findings were quite alarming to Carson, for example:

> "DDT was found in mothers' milk, in newborn infants, and in people never exposed to DDT, save for dietary contamination. Small amounts of DDT cause harm: 3.0 ppm inhibit

essential heart muscle enzymes; 5 ppm cause necrosis of liver. DDT is stored in the fat & organs of people: Individuals with no known exposure to DDT store an average of 5.3 ppm to 7.4 ppm in liver and organs; Agricultural workers: 17.1 ppm; insecticide plant workers: as high as 648 ppm."[99]

Even federal agency findings were not heeded in controlling DDT and similar substances as noted in a Federal Drug Administration Report in 1950: "It is extremely likely that the potential health hazards of DDT have been underestimated."[100]

These scientific papers were unlikely to be seen by the public, and few scientists issue press releases or summaries of their findings for the lay public. The mystique of science in the ivory tower of academia does not serve well the need for informed citizens to participate in the complex public policy process. Rachel Carson distilled the information and reflected it back in a narrative the public could understand. It was the story, the visualization of unintended consequences and inevitable harm that gave Rachel Carson her credibility and her power. Her "The Other Road" chapter that closes *Silent Spring* evokes a plea for caution, and reflection on the consequences before acting on a broad scale to introduce contaminants into the living earth.

We stand now where two roads diverge. But unlike the roads in Robert Frost's familiar poem, they are not equally fair. The road we have long been traveling is deceptively easy, a smooth superhighway on which we progress at great speed, but at its end lies disaster. The other fork in the road—the one less traveled by—offers our last, our only chance to reach a destination that assures the preservation of the Earth.[101]

Silent Spring caught the attention of President John Kennedy. He had long appreciated Rachel Carson's work from her books on the sea and adopted several coastal wildlife refuges as a result of her influence. Rachel Carson was invited to give testimony to Congress in which she set out the foundations for a regulatory protocol that would protect living systems from broadscale contamination from chemicals. Kennedy supported Carson publicly: Because of "Miss Carson's book," Kennedy said in a televised press conference, the Department of Agriculture and the Public Health Service had launched a full-blown investigation into whether pesticides caused illnesses in humans.[102]

When *Silent Spring* was published, he supported the Congressional hearings in both the Senate and the House of Representatives to investigate the issue of contamination from pesticides. Here are her principal findings and recommendations from her Testimony

> The contamination of the environment with harmful substances Is one of the major problems of modern life. The world of air and water and soil supports not only the hundreds of thousands of species of animals and plants; It supports man himself. In the past we have often chosen to ignore this fact. Now we are receiving sharp reminders that our heedless and destructive acts enter into the vast cycles of the earth and in time return to bring hazard to ourselves.[103]

Her testimony presented evidence of ways pesticides were transmitted and dispersed in air, water and soil beyond the areas treated. She made a series of detailed recommendations summarized here:

> All the foregoing evidence, it seems to me, leads inevitably to certain conclusions. The first is that aerial spraying of pesticides should be brought under strict control and should be

reduced to the minimum needed to accomplish the most essential objectives. Reduction would, of course, be opposed on the grounds of economy and efficiency. If we are ever to solve the basic problem of environmental contamination, however, we shall have to begin to count the many hidden costs of what we are doing, and weigh them against the gains or advantages.

The second conclusion that seems apparent is that a strong and unremitting effort ought to be made to reduce the use of pesticides that leave long-lasting residues, and ultimately to eliminate them. This, you will remember, was one of the recommendations of the President's Science Advisory Committee. I strongly concur in this recommendation, for I can see no other way to control the rapidly spreading contamination I have described.

There are several other recommendations I would like to suggest, bearing on various specific aspects of the immensely complex pesticide problem. These are as follows: (condensed)

> 1. Right of citizens to be informed when poisons are being sprayed in their neighborhood

> 2. Support for medical research and education about pesticides, their effects on health, and the interactions among various man-made chemicals and drugs in the body

> 3. Establish Standards for the sale and use of pesticides and poisons

> 4. Registration of chemicals and some control of their indiscriminate use

5. Investigate and promote the use of biological controls for insect management

6. Inform physicians of the impact of biocides on health

7. Protect workers who are heavily exposed, especially workers.[104]

In 1970, Congress finally acted on the impetus Rachel Carson initiated, under President Nixon. Between Rachel Carson's death in April 1964 and November 1970, America was in turmoil not only over pollution and radio-active fallout, but dramatic images of the Cuyahoga River in Ohio in flames, with destruction of bridges and buildings, a large spill of crude oil off the coast of the University of California in Santa Barbara, and all exacerbated by protests of the Vietnam War and the emergence of Women's Liberation. In these tumultuous times, it seemed politically expedient to make some demonstration of response.

The Environmental Protection Agency was established to coordinate the mission of protecting the nation's air, water and soil, and the foundational environmental protection laws were enacted within five years:[105]

- Environmental Protection Agency (EPA) established in in 1970 authority to set tolerances for chemical residues.

- Clean Air Act (1970)

- Federal Water Pollution Control Act, 1948—"Clean Water Act" significantly reorganized and expanded in 1972.

- Endangered Species Act 1973

- Congress amended Federal Insecticide, Fungicide and Rodenticide Act in 1972 to transfer pesticide regulation to the EPA and mandated protection of public and environment health.

- The EPA ceased licensing DDT in 1972.

- The Toxic Substances Control Act of 1976 was *Silent Spring's* greatest legal vindication.

Throughout her life as a scientist and as a writer, Rachel Carson kept drawing attention to the interconnected web of life and how the health of humans was inextricably connected to the health of the environment.

It took hundreds of millions of years to produce the life that now inhabits the earth – eons of time in which that developing and evolving and diversifying life reached a state of adjustment and balance with its surroundings. The environment, rigorously shaping and directing the life it supported, contained elements that were hostile as well as supporting. …Given time—time not in years but in millennia—life adjusts, and a balance has been reached. For time is the essential ingredient, but in the modern world there is no time.[106]

The body of her work is based on the precautionary principle—the need to protect the environment and natural resources for future generations. Her environmental ethic can be simply summarized in four key concepts:[107]

- Live in harmony with nature.

- Preserve and learn from natural places.

- Minimize the effects of synthetic chemicals on the natural systems of the world.

- Consider the implications of all human actions on the global web of life.

Today, we are facing overlapping and interconnected existential threats to the survival of life on Earth as we have known it.

Global Warming in 2020 already has reached

1.1°C increase over pre-industrial levels with projections of 4°C to 6°C increase by 2050 from feedback with existing emissions.[108]

Global Pollution: 95% of the global population breathes in polluted air daily.[109] Every year, about 8 million tons of plastic waste escapes into the oceans from coastal nations.[110]

Loss of ecosystems: We are eroding the very foundations of our economies, livelihoods, food security, health and quality of life worldwide. 1,000,000 species are threatened with extinction.[111]

Loss of Ice Fields: From the Arctic to Peru, from Switzerland to the equatorial glaciers of Man Jaya in Indonesia, massive ice fields, monstrous glaciers, and sea ice are disappearing, fast. Millions of gallons of freshwater flow into the oceans, slowing the great thermal-haline circulation belts.[112]

Climate refugees: Severe droughts and floods disrupt communities all over the world, especially in coastal areas and island nations.[113] UN predicts 1.2 billion climate refugees by 2050.[114]

The regulatory system established in the decade following *Silent Spring* has failed to protect the air, the water, the land, or the health of people. In its history, the E.P.A. has mandated safety testing for only 200 of the 85,000 industrial chemicals available for use today. Once chemicals are in use, the burden on the E.P.A. is so high that it has succeeded in banning or restricting only six substances, and often only in specific applications: DDT, polychlorinated biphenyls, dioxin, hexavalent chromium, asbestos and chlorofluorocarbons.[115]

The risk to people and living things is a function of the inherent hazard of the material and the amount of exposure. The approach adopted in the EPA regulatory system manages risk by controlling the exposure, setting limits on the amount of toxic materials that are considered safe to expose to the public. This approach ignores one of Rachel Carson's most central concerns—that we are subject to a chemical stew, a combination of chemicals and toxins emitted by multiple sources and mixed together in the air and water and soil, so people encounter complex mixtures of materials, individually permitted as safe.

This approach has failed to protect us. Nearly half of our rivers and streams and more than one-third of our lakes are polluted and unfit for swimming, fishing, and drinking.[116] Even more alarming is the distribution of toxic materials from superfund sites into rivers and spread over the land due to extreme flooding. Severe storms and flooding contaminate hundreds of acres as superfund sites overflow onto farmland and residential and business areas. Thirteen of the 41 Superfund sites in Texas were flooded by Harvey and were "experiencing possible damage" due to the storm.[117] The U.S. produces more than 30 million tons of hazardous waste each year.[118] U.S. put 76 million tons of pollution into the atmosphere in 2021.[119]

Much of the hazardous waste comes from the production of petrochemical products such as plastics, especially single-use disposable materials. There are now 15–51 trillion pieces of plastic in the world's oceans—from the equator to the poles, from Arctic ice sheets to the sea floor.[120] Not one square mile of surface ocean anywhere on earth is free of plastic pollution. The fossil fuel industry plans to *increase plastic production* by 40 percent over the next decade.[121] These oil giants are rapidly building petrochemical plants across the United States to turn fracked gas into plastic. This means more toxic air pollution and plastic in our oceans.

One of Rachel Carson's greatest concerns which she explained in depth to Congress, was that toxic chemicals would migrate through food chains and become a health threat to all living things, including people. This concern has been proven in the body burden of all people tested by the CDC, even infants at birth. A recent biomonitoring study by the Mount Sinai School of Medicine found

167 synthetic chemicals found in people screened; 76 are known to cause cancer in humans or animals; 94 are toxic to the brain and nervous system; 79 cause birth defects or abnormal development.[122] The dangers of exposure to these chemicals **in combination** has never been studied.[123]

Sixty years ago Rachel Carson wrote *Silent Spring* as a clarion call of warning, a plea for caution in dispersing potentially harmful synthetic chemicals into the biosphere without understanding their potential for harm. Her advice has been ignored, and we see the consequences of protecting industrial profit, apparent convenience, and profligate waste as the quality of our environment degrades day by day around us. We are not facing a technology problem. Solutions to the energy, materials, food and water problems that inundate us are readily at hand:

Shifting to renewable and sustainable energy systems; practicing regenerative agriculture to restore fertility of the ground and capture carbon in the soil; circular materials management based on non-toxic feedstocks and non-toxic byproducts through green chemistry; and preserving and respecting the dignity of people.[124] We can build a shared prosperity where people can thrive on a living planet.[125] Our life support system comes from the dynamic ecosystems of the earth that provide fresh air, clean water, fertile ground, and a biodiversity of species, of which humans are but one part. We must learn to live in harmony with nature. It is an ethical and moral choice. The laws of Nature are not negotiable. I close with Rachel Carson's words:

> Underlying all of these problems of introducing contamination into our world is the question of moral responsibility—not only to our own generation, but to those of the future. We are properly concerned for the health of those now alive; but the threat is infinitely greater to generations unborn; to those who have no voice in the decisions we make today, a heavy one.[126]

EPILOGUE

THE TRIUMPH OF LIFE RENEWED

Nurse tree in Devil's Hollow

"For all at last return to the sea—to Oceanus, the ocean river, like the overflowing stream of time, the beginning and the end."
Rachel Carson,"The Sea Around Us," p. 59

This summer of 2022 has brought the full vulnerability of life to focus. Personal health challenges especially from finally recognizing that my eyes are failing has humbled me. Now, among all the pieces of myself lost, I have plastic lenses in my eyes. My daughter is in the middle of chemotherapy for a return of her breast cancer, with hopeful prospects of success. I see my grandchildren and nieces and nephews graduating into the next phase of their adult lives, and I send them into this fractured world with trepidation, but also with hope that their own creativity and dedication can make a difference in the world they build.

It was a great pleasure on my 76th birthday to attend a utility convened seminar releasing a "Beneficial Electrification" report to see four strong young women leading the research and heading the discussion. Two were former students. I recalled my early days in the utility arena, where I was most often the only woman in the room. I was moved to see their creativity and leadership blossom.

As I visit my family in small gatherings, still under the cloud of the COVID pandemic now in multiple variants, I feel the bonds formed so many years ago crossing generations with traditions and woven with love that holds fast. Shared gardens and shared stories, embellished with new adventures, carry us all forward with a sense of connection to each other and to the past. Tom in the heavy phase of treatment for his cancer thrives on the stories he shares with his brothers and children. Family brightens the darkest times, even as we all know we are living on borrowed time, inching daily toward an unknown end.

I sit here in the garden, full of promise for a luxuriant summer, amid the unfurling iris and budding roses under cascades of honeysuckle and watch the garden birds come to and fro tending their nests. It is especially comforting to see the robins nest year after year close to my house. I wonder at their migration and whether they will return safely. For four years, I had a banded male robin, the Master of the Garden, who came to claim his territory and choose a mate. The ground here is fertile and full of earthworms and insects to feed his flock.

I heard a wood thrush trill in the early dawn today, a creature of the woods not seen here recently. The cardinals, Carolina chickadees, song sparrows, and blue jays as well as finches, Carolina wrens, crows, and hawks stay year-round. I take comfort from seeing these harbingers of the state of our world returned from their winter migrations to grace my garden. I hope that the cessation of construction at the school next door and the deliberate protection of the urban forest canopy of Forest Hills will sustain the wild songbirds for years to come.

I also see hope growing in the people of my neighboring communities as they struggle to build a more resilient and economically equitable future. Recovering industrial towns by re-investing in the people and cleaning up the mess of the old industrial past is an exciting enterprise. Eco-industrial parks are in the planning to make materials and products designed from bamboo, industrial hemp, and mushrooms rather than fossil feedstocks. Environmental justice movements abound all over the country, and even in the international space.

The general opinion and the desires of the people are becoming more aligned, even amid the partisan rifts and dis-information largely driven by those with a vested interest in preventing change. The laws of nature are not negotiable. The fossil extractive industries will fall to reality. It is only a matter of whether the vision for a finer future, a sustainable future, a healthier future can prevail before irreversible damage has occurred. I am not among those who believe it is already too late, but there is no time to waste. We live but a moment in the stream of time. Each of us has an imprint on the world we live in, on the people we interact with, and on the society we keep. As the frailty of age begins to limit physical action, the luxury of moments in good health become precious treasures to fill with thoughts captured and shared in letters to distant friends, in conversations shared with loved ones, constituents and written to the world. The ancient wisdom of the trees infuses my mind with courage to keep teaching, learning and believing in the inherent beauty of this living earth.

I think of Rachel Carson as she wrote her last essays and reflections in letters to her close friend Dorothy Freeman. I plant milkweed to draw the monarch butterflies and I plant a garden of white hyacinth in her honor. Her courage in the face of so many obstacles and her accomplishments in writing through her pain and failing health inspire me always.

As I sit weeding at the edge of my front garden, a neighbor passes with her baby stroller and we have a short chat. I hand her baby a bright yellow buttercup and relish her sweet smile. I look into the eyes of this little child, and I know the earth will be here to serve her. All our children deserve a future. The earth can heal, but we must help.

Blessed Be.

ACKNOWLEDGMENTS
AND THANKS

No words suffice to express the debt of gratitude I feel in my heart for so many people who have encouraged me on this journey. My partner, Tom Jensen, has been traveling this hardest road with me as we battle illness together and keep the simple pleasures of living intact and close. My brave, beautiful daughter, whose life struggles and now shared battle with breast cancer, amazes me with her creativity, courage and strength. To my son, Steve, I am forever grateful for the endless gift of a positive mental attitude and shared joy in the beauty of living each moment; my brother, Michael, and sister-in law, Maryann; my sister, Linda, and her husband, Randy; all my cousins and vast extended family have offered encouragement and support, anecdotes remembered from past times, and hours of shared memories.

I gratefully acknowledge the many colleagues whose work has enhanced and amplified my own, especially Mark Dixon, producer of *The Power of One Voice* film; Kirsi Jansa, producer of our *Sustainability Pioneers* videos; Ann Rosenthal, whose artist's eye offers perspective and reflection; David Carlisle and Bob Musil, who gave me courage to put this journey into words. To my fellow warriors in the battle to make this world a healthier happier place, I extend my gratitude and my debt of honor to Matt Mehalik, Dianne Peterson, Terry Collins, Arlene Blum and Amanda Woodrum. To all my friends from the Battle of Homestead Foundation who encouraged me to write this story, especially Rosemary Trump, Charlie McCollester, Steffi Domike, Mike Stout, Dee Korchirka and Wanda Guthrie. I thank my colleagues in the Phipps Conservatory and Botanical Gardens where I take refuge when I need inspiration, especially Richard Piacentini, Greg Dufour and Sarah States.

I am humbled by the many people whose ideas, shared stories, and encouragement brought this little volume to light at last: Maren Cook; PK Weston; my dear friends, Caroline A. Mitchell and Diane Law; my colleagues, Barbara Martin and Maureen Puskar; my neighbors and friends, Nancy Ellis, Linda Hyde, Phoebe Marshall and Gail Harper. And truly to Linda Lear whose mentorship and wisdom kept my hope alive for many years.

My students give me the motivation to keep on striving to make this world a better place. They number in the hundreds now, and I still see their faces—even in the little boxes on Zoom. I see their accomplishments with pride and send them my very deepest gratitude for their strength and their passion. Their hopes and aspirations deserve to be realized as the true vision for the future shines in their eyes. I hope that in a small way I have shared with them the inspiration of Rachel Carson's mighty ethic to sustain them forward.

And last of all, I thank John Stanko, my editor and publisher with Urban Press for bringing this book to light. His insight and confidence in my work has carried this finally to fruition.

ENDNOTES

Endnotes to Part 1

[1]Linda J. Lear. *Rachel Carson: Witness for Nature*. Henry Holt & Company. New York. 1997. Page 461.

[2]Linda Lear (Ed.) Rachel Carson. 1963 Letter to Dr. George Crile, Jr. *Lost Woods: The Discovered Writing of Rachel Carson*. Beacon Press. Boston 1998. Page 223-226.

[3]Kevin Hillstrom. " Tecumseh Decries Land Cessions: White People Have no Right to Take the Land from the Indians." *U.S. Environmental Policy and Politics: A Documentary His*tory. CQ Press. Washington D.C. 2010 Page 57-58.

[4]Meredith Wheeler. Operational Group PAT: A Fresh Look. July 4, 2011. http://www.ossreborn.com/files/OG_PAT_A_Fresh_LookPhotos1.pdf

Endnotes to Part 3.

[5]U.S. Breast Cancer Statistics http://www.breastcancer.org/symptoms/understand_bc/statistics
Accessed September 20, 2022.

[6]Marc Hafsted. "Do Environmental Regulations Kill Jobs?" Resources. August 14, 2019. https://www.resources.org/common-resources/do-environmental-regulations-kill-jobs/ Accessed September 20, 2022.

[7]Centers for Disease Control and Prevention, National Biomonitoring Program. https://www.cdc.gov/biomonitoring/index.html Accessed September 20, 2022.

[8]Edna Ribiero et. Al. Occupational Exposure to Bisphenol A (BPA): A Reality That Still Needs to Be Unveiled. Toxics. 2017 Sep; 5(3): 22. Published online 2017 Sep 13. https://www.ncbi.nlm.nih.gov/pmc/articles/PMC5634705/ Accessed September 20, 2022.

[9]Sara Goodman. "Tests Find More Than 200 Chemicals in Newborn Umbilical Cord Blood." *Scientific American*. December 2, 2009. https://www.scientificamerican.com/article/newborn-babies-chemicals-exposure-bpa Accessed September 20, 2022.

[10]Joe McCarthy. "9 Shocking Facts About Plastics in Our Oceans." *Global Citizen*. June 12, 2017. https://www.ecowatch.com/plastic-oceans-facts-images-2436857254.html Accessed September 20, 2022.

[11]James. How Does The German Pfand System Work, And Is It

Effective? 21 May 2017. https://liveworkgermany.com/2017/05/how-does-the-german-pfand-system-work-and-is-it-effective/ Accessed September 20, 2022.

[2]Beth Terry. 100 Steps to a Plastic Free Life. https://myplasticfreelife.com/plasticfreeguide/ Accessed September 20, 2022.

[13]National Council of State Legislatures offers resources and model legislation. HTTP://WWW.NCSL.ORG/RESEARCH/ENVIRONMENT-AND-NATURAL-RESOURCES/PLASTIC-BAG-LEGISLATION.ASPX Accessed September 20, 2022.

[14]Environmental Protection Agency.Toxic Release Inventory. National Analysis 2015. www.epa.gov Accessed September 20, 2022.

[15]Patricia M. DeMarco. *Pathways to Our Sustainable Future: A Global Perspective from Pittsburgh*. (University of Pittsburgh Press, Pittsburgh. 2017) Pages 140-169.

[16]World Meteorological Organization. Air Quality and Climate Bulletin No. 2 -September 2022. https://public.wmo.int/en/our-mandate/focus-areas/environment/air_quality/wmo-air-quality-and-climate-bulletin-no.2 Accessed September 29, 2022.

[17]Monica Anderson. For Earth Day 2017, here's how Americans view environmental issues." Pew Research Center. April 20, 2017. http://.www.pewresearch.org/fact-tank/2017/04/20/for-earth-day-heres-how-americans-view-environmental-issues/ Accessed September 22, 2022.

[18]Rachel L. Carson. *Silent Spring*. Houghton Mifflin Company, Boston,1962. Page 6.

[19]White House, Briefings and Statements. Energy and Environment Archive. 2017-2018 https://www.whitehouse.gov/briefings-statements/?issue_filter=energy-environment

[20]Juliana vs.The United States. Constitutional Climate Lawsuit filed in U.S. District Court in the District Court of Oregon. 2015. https://www.ourchildrenstrust.org/us/federal-lawsuit/

[21]Grant Crawford. "Tri-Council Passes Resolution supporting Standing Rock Sioux."Talequah Daily Press. May 1, 2017. http://www.tahlequahdailypress.com/news/tri-council-passes-resolution-supporting-standing-rock-sioux/article_89c0d220-2e88-11e7-9633-17825b450097.html

[22]Daniel Walmer. "PA Senator wants protestors to cover costs if they break the law." Lebanon Daily News.August 26, 2017. https://www.ldnews.com/story/news/local/2017/08/26/pa-senator-wants-protesters-cover-costs-if-they-break-law/601452001/

American Legislative Exchange Council. "Model Policy: Critical

Infrastructure Protection Act" https://www.alec.org/model-policy/critical-infrastructure-protection-act/ (Under consideration in nine states, including Pennsylvania.)

[23]Portions of this statement were developed in collaboration with Mike Stout, Anita Prizio, Jay Ting Walker, Cole McDonald, with input from Jules Lobel and Mark Dixon as part of a proposed Platform for the Community Power Movement.

[24]Constitution of the Commonwealth of Pennsylvania, Article 1, Section 27.

[25]World Peoples Conference on Climate Change and the Rights of Mother Earth. Universal Declaration of the Rights of Mother Earth. Cochabamba, Bolivia. April 22, 2010. https://therightsofnature.org/universal-declaration/

[26]First National People of Color Environmental Leadership Summit, "Principles of Environmental Justice." Washington, D.C. October 27-29, 1991. https://www.ejnet.org/ej/principles.html

[27]Intergovernmental Panel on Climate Change, *Climate Change 2007: Synthesis Report*, Contribution of Working Groups I, II, and III to the Fourth Assessment Report of the Intergovernmental Panel on Climate Change, IPCC Plenary 27, Valencia, Spain, November 12-17, 2007, page 36.

[28]Crimmins, A., J. Balbus, J.L. Gamble, C.B. Beard, J.E. Bell, D. Dodgen, R.J. Eisen, N. Fann, M.D. Hawkins, S.C. Herring, L. Jantarasami, D.M. Mills, S. Saha, M.C. Sarofim, J. Trtanj, and L. Ziska, Eds. *The Impacts of Climate Change on Human Health in the United States: A Scientific* Assessment. U.S. Global Change Research Program, Washington, DC, 312 pp. http://dx.doi.org/10.7930/J0R49NQX https://health2016.globalchange.gov:

[29]Centers for Disease Control and Prevention. "Vital Signs: Asthma in the United States." May 2011. https://www.cdc.gov/vitalsigns/asthma/index.html

[30]National Institute of Health, National Cancer Institute. Cancer Statistics. https://www.cancer.gov/about-cancer/understanding/statistics

[31]Sara Goodman. "Tests find more than 200 chemicals in newborn umbilical cord blood." Scientific American. December 2009. https://www.scientificamerican.com/article/newborn-babies-chemicals-exposure-bpa/

[32]Hagai Levine Niels Jørgensen Anderson Martino-Andrade Jaime Mendiola Dan Weksler-Derri Irina Mindlis Rachel Pinotti Shanna H Swan Temporal trends in sperm count: a systematic review and meta-regression analysis. *Human Reproduction Update*, Volume 23,

Issue 6, 1 November 2017, Pages 646–659, https://doi.org/10.1093/humupd/dmx022

[33]Matthew Mehalik, Executive Director, The Breathe Project contributed to this discussion.

Endnotes for Chapter 4.1: Lessons from the Standing Rock Sioux

[34]Charels McCollester. *The Point of Pittsburgh*. The Battle of Homestead Foundation. Pittsburgh. 2008. Pages 9-12.5

[35]Kristina Marusic. "Fracking conference and opposing tribal rally highlight competing visions for the future of Western Pennsylvania." The Daily Climate Newsletter. October 23, 2018. https://www.daily-climate.org/shale-insight-convention-native-protest-2614507301/we-were-forced-off-this-land-but-were-lucky-to-remain-in-the-vicinity-of-the-allegheny-river

[36]Ibid

[37]Rachel Carson. *Silent Spring*. Houghton Mifflin Company: Boston 1962, p. 277.

Endnotes for Chapter 4.2: Lessons from the Hibakusha

[38]Dennis Normile. "How atomic bomb survivors have transformed our understanding of radiation's impacts." Science. July 23, 2020. https://www.sciencemag.org/news/2020/07/how-atomic-bomb-survivors-have-transformed-our-understanding-radiation-s-impacts Accessed August 5, 2020.

[39]Harry S. Truman. August 6, 1945: Statement by the President Announcing the use of the A-Bomb at Hiroshima. Presidential Speeches. University of Virginia, Miller Center. https://millercenter.org/the-presidency/presidential-speeches/august-6-1945-statement-president-announcing-use-bomb

[40]Health Risks of Radon and Other Internally Deposited Alpha-Emitters. Beir IV. National Research Council (US) Committee on the Biological Effects of Ionizing Radiations. Washington (DC): National Academies Press (US); 1988. https://www.ncbi.nlm.nih.gov/books/NBK218114/

[41]Health Risks of Radon and Other Internally Deposited Alpha-Emitters. Beir IV. Table 7-1 Transuranium Nuclides of Potential Biological Significance. National Research Council (US) Committee on the Biological Effects of Ionizing Radiations. Washington (DC):

National Academies Press (US); 1988. https://www.ncbi.nlm.nih.gov/books/NBK218114/

[42] *The Nevada Test Site*. Emmet Gowin. Foreword by Robert Adams. Princeton and Oxford: Princeton University Press, 2019, pages 148 and 157 (Publ. DOE/NV-209, 1993).

[43] Johnson, Carl (1984). "Cancer Incidence in an Area of Radioactive Fallout Downwind From the Nevada Test Site". *Journal of the American Medical Association*. **251** (2): 230. doi:10.1001/jama.1984.03340260034023

[44] *"Hanford Site: Hanford Overview"*. *United States Department of Energy.*

[45] Deutsch, William J.; et al. (2007). *Hanford Tanks 241-C-202 and 241-C-203 Residual Waste Contaminant Release Models and Supporting Data*. Pacific Northwest National Laboratory (PNNL). doi:10.2172/917218

[46] **Address by Mr. Dwight D. Eisenhower, President of the United States of America, to the 470th Plenary Meeting of the United Nations General Assembly.** *Tuesday, 8 December 1953.* https://www.iaea.org/about/history/atoms-for-peace-speech

[47] *Strauss, Lewis (16 September 1954). Remarks prepared by Lewis L. Strauss (PDF) (Technical Report). United States Atomic Energy Commission.* https://www.nrc.gov/docs/ML1613/ML16131A120.pdf

[48] Source: U.S. Energy Information Administration, International Energy Statistics, as of April 16, 2020 https://www.eia.gov/energy-explained/nuclear/nuclear-power-plants.php

[49] Institute of Medicine and National Research Council. 2007. Advancing Nuclear Medicine Through Innovation. Washington, DC: The National Academies Press. https://www.ncbi.nlm.nih.gov/books/NBK11471/

[50] Statement by Lassina Zerbo, Executive Secretary, Comprehensive Nuclear Test-Ban Treaty Organization (CTBTO)Vienna, 21 April 2018. https://www.ctbto.org/press-centre/press-releases/2018/statement-by-lassina-zerbo-executive-secretary-comprehensive-nuclear-test-ban-treaty-organization-ctbto/

[51] For an overview of treaties and Agreements on nuclear matters see https://www.armscontrol.org/treaties

Endnotes for Chapter 4.3:
Re-Imagine America in Harmony with Nature

[53]Rabah Kamal, Nisha Kurani, Daniel McDermott, and Cynthia Cox. "How prepared is the U.S. to respond to COVID-19 relative to other countries?" Petersen-Kaiser Foundation. Health System Trakcer. March 27, 2020. https://www.healthsystemtracker. org/?sfid=4356&_sft_category=access-affordability,health-well-being,spending,quality-of-care Accessed April 12, 2020.

[54]National Research Council (US); Institute of Medicine (US); Woolf SH, Aron L, editors. *U.S. Health in International Perspective: Shorter Lives, Poorer Health.* Washington (DC): National Academies Press (US); 2013. Summary. Available from: https://www.ncbi.nlm.nih. gov/books/NBK154469/

[55]Edward R. Berchick, Jessica C. Barnett, and Rachel D. Upton. "Health Insurance Coverage in the United States: 2018" United States Census Bureau. Report Number P60-267 (RV). November 8, 2019. https://www.census.gov/library/publications/2019/demo/ p60-267.html Accessed April 12, 2020

[56]https://www.cdc.gov/reproductivehealth/maternalinfanthealth/infantmortality.htm

[57]Alex Graff. "Women of Color Far More Likely to Die in US from Pregnancy-Related Causes" The Globe Post, National Newsletter. November 21, 2019.

https://theglobepost.com/2019/11/21/us-maternity-infant-mortality/ Accessed 7.Feb. 2020

[58]National Research Council (US); Institute of Medicine (US); Woolf SH, Aron L, editors. U.S. Health in International Perspective: Shorter Lives, Poorer Health. Washington (DC): National Academies Press (US); 2013. Summary. Available from: https://www.ncbi.nlm.nih. gov/books/NBK154469/

[59]Karen Davis, Kristof Stremikis, David Squires, and Cathy Schoen. "How the U.S. Health Care System Compares Internationally." The Commonwealth Fund. June 16, 2014. https://www.commonwealthfund.org/publications/fund-reports/2014/jun/mirror-mirror-wall-2014-update-how-us-health-care-system Accessed April 12, 2020.

[60]Brad Plummer and Nadja Popovich. "The World Still Isn't Meeting Its Climate Goals." The New York Times. December 7, 2018. https:// www.nytimes.com/interactive/2018/12/07/climate/world-emissions-paris-goals-not-on-track.html Accessed April 3, 2020.

[61]Christopher Ingraham. "Air Pollution is getting worse and data show

more people are dying."Washington Post. October 29, 2019. https://www.washingtonpost.com/business/2019/10/23/air-pollution-is-getting-worse-data-show-more-people-are-dying/ Accessed April 3, 2020.

[62]Karen Clay, Nicholas Z. Miller. "Recent Increases in Air Pollution: Evidence and Implications for Mortality." National Bureau of Economic Research, Working Paper No. 26381. October 2019. https://www.nber.org/papers/w26381 Accessed April 13, 2020.

[63]Karen Clay and Nicholas Z. Miller. "Recent Increases in Air Pollution: Evidence and Implications for Mortality." The National Bureau of Economic Research. October 2019. NBER Working Paper No 26381. Quoted in https://www.washingtonpost.com/business/2019/10/23/air-pollution-is-getting-worse-data-show-more-people-are-dying/ Accessed April 3, 2020.

[64]Congressional Budget Office, "Projected Changes in the Distribution of Household Income, 2016-2021," December 2019, Figure 4, p. 15, https://www.cbo.gov/publication/55941. Accessed April 12, 2020.

[65]Chad Stone, Danilo Trisi, Arloc Sherman and Jennifer Beltran. A Guide to Statistics on Historical Trends in Income Inequality. Center on Budget and Policy Priorities. January 13, 2020. https://www.cbpp.org/research/poverty-and-inequality/a-guide-to-statistics-on-historical-trends-in-income-inequality Accessed April 12, 2020.

[66]Urban Institute calculations from Survey of Financial Characteristics of Consumers 1962 (December 31), Survey of Changes in Family Finances 1963, and Survey of Consumer Finances 1983–2016. Notes: 2016 dollars. No comparable data are available between 1963 and 1983.

http://apps.urban.org/features/wealth-inequality-charts/

[67]Thomas Berry. "The Ecozoic Era." Eleventh Annual E.F. Schumacher lecture. Schumacher Center for a New Economics. October 19, 1991. https://centerforneweconomics.org/publications/the-ecozoic-era/ Accessed April 12, 2020.

[68]L. Hunter Lovins et.al. *A Finer Future – Creating an Economy in Service to Life*. New Society Publishers. B.C. Canada. 2018. Page 27

[69]Universal Declaration of Rights of Mother Earth. World People's Conference on Climate Change and the Rights of Mother Earth. Cochabamba, Bolivia. April 22, 2010. https://therightsofnature.org/universal-declaration/

[70]Gustafson, A., Goldberg, M. H., Kotcher, J. E., Rosenthal, S. A., Maibach, E. W., Ballew, M. T., & Leiserowitz, A. (2020). Republicans and Democrats differ in why they support renewable energy. Energy Policy, 141. DOI: 10.1016/j.enpol.2020.111448 https://climatecommunication.yale.edu/publications/republicans-

and-democrats-differ-in-their-primary-reasons-for-supporting-renewable-energy/ Accessed April 13, 2020.

[71]For detailed discussions of implementation for these pathways see: Patricia DeMarco. *Pathways to Our Sustainable Future*. University of Pittsburgh Press. Pittsburgh. 2017

[72]Mark Dixon, Andre Goes, Heather Haar, Joanne Martin, Connor Mulvaney and Sophie Reidel. *Re-Imagine Beaver County*. League of Women Voters. Spring 2019. https://www.shalepalwv.org/wp-content/uploads/2019/05/2019-Re-Imagine-Beaver-County-Book.pdf Accessed September 20, 2022

[73]Carl Zimmer. E. O. Wilson, A Pioneer of Evolutionary Biology, Dies at 92." The New York Times. December 27, 2021. https://www.nytimes.com/2021/12/27/science/eo-wilson-dead.html Accessed September 20, 2022.

[74]United Nations Environment Programme. Why do Sustainable Development Goals Matter? Goal 2:Zero Hunger. https://www.unep.org/explore-topics/sustainable-development-goals/why-do-sustainable-development-goals-matter/goal-2

[75]United Nations Environment Programme> Rethinking Food Systems. June 4, 2021. https://www.unep.org/news-and-stories/story/rethinking-food-systems

[76]Jack Uldrich. Regenerative Agrculture: The Next Trend in Food Retailing. Forbes. August 19, 2021. https://www.forbes.com/sites/forbesbusinesscouncil/2021/08/19/regenerative-agriculture-the-next-trend-in-food-retailing/?sh=513fef962153 Accessed September 20, 2022.

[77]Oliver Morrison. "Regenerative Agriculture Key to Future Proofing Food Sector." FOOD: U.S. Edition. May 14, 2021. https://www.foodnavigator.com/Article/2021/05/14/Regenerative-agriculture-key-to-future-proofing-food-sector-says-EIT-Food

[78]Analysis: Was 2018 the year of the influential sustainable consumer? Nielsen IQ Insights. December 17, 2018. https://nielseniq.com/global/en/insights/analysis/2018/was-2018-the-year-of-the-influential-sustainable-consumer/

[79]Rachel Pope. "Recent Study Reveals More Than a Third of Global Consumers Are Willing to Pay More for Sustainability as Demand Grows for Environmentally-Friendly Alternatives" Business Wire. October 14, 2021. https://www.businesswire.com/news/home/20211014005090/en/Recent-Study-Reveals-More-Than-a-Third-of-Global-Consumers-Are-Willing-to-Pay-More-for-Sustainability-as-Demand-Grows-for-Environmentally-Friendly-Alternatives

[80]Beaulac J, Kristjansson E, Cummins S. A systematic review of food deserts, 1966-2007. Prev Chronic Dis 2009;6(3):A105. http://www.cdc.gov/pcd/issues/2009/jul/08_0163.htm. Accessed September 20, 2022.

[81]News WGAL: How Urban Farms Are Changing the Landscape of Food Deserts. Broadcast 4:23 PM EDT. June 24, 2021. https://www.wgal.com/article/urban-farms-food-deserts/36845521#

[82]Andrew Moore. "Building a Case for Community Gardens" North Carolina State University News. June 1, 2021. https://cnr.ncsu.edu/news/2021/06/community-gardens/

[83]Lauren Rothschild. "U.S. "Digital Divide": How internet access disparities affect resilience." Global Resource Institute, Northeastern University. https://globalresilience.northeastern.edu/us-digital-divide-how-internet-access-disparities-affect-resilience/ Accessed September 20, 2022.

[84]Rebecca Harrington. "Grass takes up 2% of the land in the United States." Business Insider. February 19, 2016. https://www.businessinsider.com/americas-biggest-crop-is-grass-2016-2

Endnotes to 4.5 A Perspective on Silent Spring after 60 Years.

[85]Rachel Carson "On the Pollution of Our Environment" presented in 1963. Linda J. Lear (Ed.) *Lost Woods: The Discovered Writing of Rachel Carson*. Beacon Press. 1998. P. 228.

[86]Rachel L. Carson. *Silent Spring*. Houghton Mifflin Co. Boston. 1962. Epigraph.

[87]Rachel L. Carson. *The Sense of Wonder*. Harper Collins. New York. 1998. P. 100.

[88]Linda Lear. *Rachel Carson Witness for Nature*. Henry Holt & Company, LLC. New York. Page 7.

[89]Linda J. Lear. *Rachel Carson Witness for Nature*. Henry Holy & Company, New York. 1997. Page 81-83.

[90]Rachel Carson Testimony. Interagency Coordination In Environmental Hazards. Pursuant to S. Res 27, 88th Congress, as amended. Tuesday, June 4, 1963 before the U.S. Senate Subcommittee on Reorganization and International Organizations of the Committee on Government Operations. Washington D.C. Senator Abraham Ribicoff presiding

[91]Rachel Carson. "Guarding our Natural resources" Conservation in Action No.5. U.S. Government Printing Office Washington DC. 1948

[92]Rachel L. Carson. *The Sense of Wonder.* Harper Collins. New York. 1998. P.131.

[93]DeMarco personal communication with E.O. Wilson. Rachel Carson Legacy Conference on Biodiversity, September 30, 2007.

[94]Wes Jackson. *Fatal Harvest: The Tragedy of Industrial Agriculture.* Land Institute. 2002.

[95]Rachel Carson. *Silent Spring.* Houghton Mifflin. Boston. 1962. p. 12.

[96]Eric Sevareid, et. Al. CBS Reports: The Silent Spring of Rachel Carson. April 13, 1963. https://www.paleycenter.org/collection/item/?q=cbs&p=34&item=T77:0033 Accessed September 18, 2022.

[97]Michael B. Smith. "Silence Miss Carson" Science, Gender and the Reception of Silent Spring." *Feminist Studies.* 2001. Vol. 27, No. 3. Pages 262-263.

[98]Paul Brooks. *The House of Life: Rachel Carson at Work.* Houghton Mifflin Company. Boston. 1989. p. 297

[99]Weyland J. Hayes, AMA Archives of Indus Health **18**:398-406. 1958

[100]Federal Drug Administration report 1950

[101]Rachel Carson. *Silent Spring.* Houghton Mifflin. Boston. 1962. Page 277

[102]Douglas Brinkley. "Rachel Carson and JFK, an Environmental Tag Team." *Audubon Magazine.* May-June 2012. https://www.audubon.org/magazine/may-june-2012/rachel-carson-and-jfk-environmental-tag-team Accessed September 15, 2022.

[103]Rachel Carson Testimony. Interagency Coordination In Environmental Hazards. Pursuant to S. Res 27, 88th Congress, as amended. Tuesday, June 4, 1963 before the U.S. Senate Subcommittee on Reorganization and International Organizations of the Committee on Government Operations. Washington D.C. Senator Abraham Ribicoff presiding.

[104]Ibid.

[105]The Origins of the EPA. https://www.epa.gov/history/origins-epa Accessed September 18, 2022.

[106]Rachel Carson. Silent Spring. Houghton Mifflin. Boston. 1962. Page 6.

[107]Patricia DeMarco. "Rachel Carson's environmental ethic – a guide for global systems decision making." *Journal of Cleaner Production.* 2017. Vol. 140. Pages 127-133. https://pubag.nal.usda.gov/catalog/5469718

[108]Daniel Glick. "The Big Thaw: As the climate warms, how much and how quickly, will Earth's glaciers melt? National Geographic.

https://www.nationalgeographic.com/environment/article/big-thaw Accessed September 18, 2022.

[109]Gavin Shaddick, Matthew L. Thomas, Heresh Amini, David Broday. Aaron Cohen, Joseph Frostad. Amelia Green. Sophie Gumy. Yang Liu, Randall V. Martin, Annette Pruss-Ustun, Daniel Simpson, Aaron van Donkelaar, and Michael Brauer. "Data Integration for the Assessment of Population Exposure to Ambient Air Pollution for Global Burden of Disease Assessment" *Environ. Sci. Technol.* 2018, 52, 16, 9069–9078 Publication Date: June 29, 2018 https://doi.org/10.1021/acs.est.8b02864 Copyright © 2018 American Chemical Society Accessed September 15, 2022.

[110]Marta Fava. Ocean Plastic Pollution an Overview: data and statistics. UNESCO. https://oceanliteracy.unesco.org/plastic-pollution-ocean/ Accessed September 17, 2022.

[111]United Nations. Nature's Dangerous Decline Unprecedented; Species Extinction Rates Accelerating. https://www.un.org/sustainabledevelopment/blog/2019/05/nature-decline-unprecedented-report/

[112]National Ocean Service. "The Global Conveyor Belt." National Oceanic and Atmospheric Administration. .https://oceanservice.noaa.gov/education/tutorial_currents/05conveyor2.html

[113]NASA. "Amazing Earth: Satellite Images from 2019." January 10, 2020. https://www.nasa.gov/feature/amazing-earth-satellite-images-from-2019

[114]Institute for Economics & Peace. Ecological Threat Report 2022: Analysing Ecological Threats, Resilience & Peace, Sydney, October 2022. Page 42. Available from: http://visionofhumanity.org/resources (accessed November 1, 2022.)

[115]Chemicals banned by EPA. https://www.epa.gov/chemicals-under-tsca

[116]U.S. EPA. National Water Quality Inventory: Report to Congress. August 2017. EPA-841-R-16-011. https://www.epa.gov/sites/default/files/2017-12/documents/305brtc_finalowow_08302017.pdf Accessed September 15, 2022.

[117]Michael Biesecker and Jason Dearen. "Toxic waste sites flooded in Houston area." The Nation. September 3, 2017. https://www.pbs.org/newshour/nation/ap-exclusive-toxic-waste-sites-flooded-houston-area Accessed September 16, 2022.

[118]U.S. Environmental Protection Agency. Report on the Environment: Hazardous Waste. 2020. https://www.epa.gov/roe/

[119]Environmental Protection Agency. National Air Quality Trends 1980-2021. https://www.epa.gov/air-trends/air-quality-national-summary

[120]Center for Biological Diversity. Ocean Plastics Pollution: A Global Tragedy for our Oceans and Sea Life. https://www.biologicaldiversity.org/campaigns/ocean_plastics/ Accessed September 18, 2022.

[121]Katie Brigham. "How the fossil fuel industry is pushing plastics on the world." CNBC. February1, 2022. https://www.cnbc.com/2022/01/29/how-the-fossil-fuel-industry-is-pushing-plastics-on-the-world-.html

Accessed September 17, 2022.

[122]Mount Sinai School of Medicine in New York, in collaboration with the Environmental Working Group and Commonweal. "The Pollution in People." EWG Report. June 14, 2016. https://www.ewg.org/research/pollution-people Accessed September 18, 2022.

[123]Linda S. Birnbaum. "State of the Science of Endocrine Disruptors." Environmental Health Perspectives. Vol .121, No.4. April 2013. https://ehp.niehs.nih.gov/doi/10.1289/ehp.1306695 Accessed September 15, 2022.

[124]Patricia M. DeMarco. *Pathways to Our Sustainable Future: A Global Perspective from Pittsburgh.* 2017 University of Pittsburgh Press. Pittsburgh..

[125]ReImagine Appalachia Blueprint. September 2021. https://reimagineappalachia.org/portfolio/the-blueprint/

[126]Rachel Carson "On the Pollution of Our Environment" presented in 1963. Linda J. Lear (Ed.) *Lost Woods: The Discovered Writing of Rachel Carson.* Beacon Press. 1998. p 242.

You can contact the author at
demarcop6@gmail.com

or stay in touch with her world at
https://patriciademarco.com

Her other book,
*Pathway to Our Sustainable Future: A Global
Perspective from Pittsburgh*
is available through the University of Pittsburgh
Press or through Amazon.